Thomas Holmes Blakesley

Papers on alternating currents of electricity for the use of students and engineers

Second Edition

Thomas Holmes Blakesley

Papers on alternating currents of electricity for the use of students and engineers
Second Edition

ISBN/EAN: 9783337214708

Printed in Europe, USA, Canada, Australia, Japan

Cover: Foto ©berggeist007 / pixelio.de

More available books at **www.hansebooks.com**

PAPERS ON
ALTERNATING CURRENTS
OF
ELECTRICITY

FOR THE USE OF

STUDENTS AND ENGINEERS

BY

T. H. BLAKESLEY, M.A.

King's' College, Cambridge,
Member of the Physical Society of London,
M. Inst. C.E.

SECOND EDITION, ENLARGED

LONDON
WHITTAKER & CO., WHITE HART ST., PATERNOSTER SQ.
AND
GEORGE BELL & SONS
1889.

BUTLER & TANNER,
THE SELWOOD PRINTING WORKS,
FROME, AND LONDON.

PREFACE.

The following chapters were written to exemplify the use of the geometrical method in treating problems involving the flow of electricity arising from the existence of sources of electro-motive force whose intensity undergoes harmonic variation.

The greater number of them appeared originally in the *Electrician* newspaper, others have appeared in the *Transactions* of the Physical Society, and in the *Philosophical Magazine*.

That on Condenser-Transformers has not been previously published.

THOMAS H. BLAKESLEY.

ROYAL NAVAL COLLEGE, GREENWICH,
 May, 1889.

CONTENTS.

CHAPTER I.
Self-Induction 1

CHAPTER II.
Mutual-Induction 15

CHAPTER III.
Condensers 23

CHAPTER IV.
Condenser in Circuit 30

CHAPTER V.
Several Condensers 37

CHAPTER VI.
Combination of Condensers with Self-Induction . 43

CHAPTER VII.
Condenser Transformer 53

CHAPTER VIII.
Distributed Condenser 58

CHAPTER IX.
Distributed Condenser (*cont.*)—Telephony . 72

CHAPTER X.

The Transmission of Power . . . 78

CHAPTER XI.

Upon the Use of the Two-Coil Dynamometer with Alternating Currents 97

CHAPTER XII.

Silence in a Telephone 108

CHAPTER XIII.

On Magnetic Lag 115

ALTERNATING CURRENTS.

CHAPTER I.

SELF-INDUCTION.

It is often taken for granted that the simple form of Ohm's law—total E.M.F. ÷ total resistance = total current—is true for alternating currents. That is to say, the E.M.F. employed in the formula is taken to be the sum of the *impressed* E.M.F.'s alone. That there are causes which modify the value of the current as deduced from this simple equation, such as mutual or self-induction, or the action of condensers, is often acknowledged in text books, and the values and laws of variation of the current are correctly stated for certain cases of instantaneous contact and breaking of circuit. But the effect of an alternating E.M.F. upon a circuit affected by self-induction, mutual-induction, and condensing action, has not been, as far as I know, put into a tangible working form.* I propose to deal with the case where the impressed E.M.F. is subject to simple harmonic variation—that is to say, the sort of variation

---

* Dr. Hopkinson has quite recently dealt with some of the cases of alternating currents in a paper read before the Society of Telegraph Engineers and Electricians.

which takes place in the apparent distance of a satellite from its primary, as seen by an eye situated in the plane of revolution, and at an indefinitely great distance from the orbit. The form of the law is simply $x = b \sin\left(\frac{\pi}{T}\overline{t-t'}\right)$, where t is the symbol for time, t' being the value at the particular epoch when x is zero, x representing the magnitude subject to the variation under consideration. Since the sine of no angle can exceed unity, b is the maximum value of x. T is half the period or the time required for x to vary from its greatest positive value to its greatest negative.

This sort of variation is exactly that which the E.M.F. undergoes in the case of a coil, turned round any axis at a uniform rate in a uniform magnetic field, and is very approximately the case in a large number of motions occurring in practice. It is also the case when a small bar magnet revolves at the centre of a coil, exactly when the magnet is very small, compared with the coil, and highly approximately in other cases. The variation of telephonic currents, as produced in a Bell telephone, is also harmonic.

Let a straight line of fixed length, and situated in the plane of the paper, undergo uniform rotation in that plane. Then its projection upon a fixed indefinitely long line also in that plane will undergo harmonic variation, and may represent any magnitude capable of undergoing such change (*e.g.* an electromotive force), the maximum value of this varying magnitude being represented by the revolving line itself. The period in which the revolving line makes one complete revolution is the period of the change. Hence, if we know the position of the fixed line and of the revolving line at any instant, we can say in what particular phase the magnitude undergoing harmonic change is at that instant. For instance, suppose these lines make 30° with each other, we can say at once that the magnitude is removed from its maximum value by an interval of time equal to one twelfth of the period. If the angle is at the instant increasing, the magnitude has passed its maximum value that interval of time ago. If the angle is growing less, the magnitude will attain its maximum after that interval of time. It is therefore necessary to fix a positive direction of rotation as representing the positive lapse of time. [That direction which is opposite to that of the hands of a watch will here be adopted.]

It follows that when we have two such electromotive forces acting in the same circuit, having different maximum values but the same Period, since each is represented by the projection of a revolving straight line upon a fixed straight line, the resultant electromotive force at the instant is the algebraical sum of the individual projections. And if the two revolving lines are laid down as the two sides of a triangle *taken in order*, the rotation being uniform and the same for both lines, the lines will remain always inclined at the same angle to each other, and the algebraical sum of their projections is the projection of the third side. Thus, in the matter of such electromotive forces, we have a theorem exactly corresponding to the triangle of directed quantities.

We may extend this mode of representing such quantities so as to form a theorem corresponding to the polygon of directed quantities, and cite it thus:—

"If the straight lines AB, BC, CD,...ST represent the maximum values of different electromotive forces, and, as to direction, are so laid down upon the paper that their projections upon a fixed straight line represent at some point of time the instantaneous values of

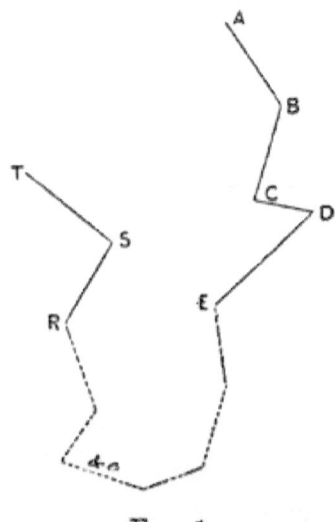

Fig. 1.

those electromotive forces, their instantaneous resultant is the projection of the simple straight line AT."

If, in any particular case, we have taken into consideration all the electromotive forces concerned, then clearly the line representing the resultant corresponds in phase with the instantaneous current; and if by scaling or calculation we find the value of this resultant in volts, we have only to divide by the resistance in ohms to obtain the maximum value of the alternating current resulting from all the component electromotive forces. This is true, even if one of the electromotive forces is that of self-induction. But suppose we have arrived at a preliminary resultant by compounding all the electromotive forces with the exception of that of self-induction; we then require the final resultant, and we obtain it by remembering that it must be at right angles to the electromotive force of self-induction; for the electromotive force of self-induction must be greatest when the current is passing through zero: therefore it must have its projection on the fixed line greatest when that of the final resultant (corresponding with the current) is zero. Therefore the final resultant and the electromotive force of self-induction must be to the preliminary resultant as the two sides of a right-angled triangle including the right angle, are to the hypothenuse; and as we already possess the hypothenuse, we have only to determine the ratio of the sides, and upon which side of the hypothenuse they must be placed, in order fully to determine the position and size of the final resultant and the electromotive force of self-induction. The geometrical construction is as follows:—

From one end of the preliminary resultant set off an angle in the negative direction of rotation, whose tangent is equal to the product of the coefficient of self-induction and the angular velocity of rotation divided

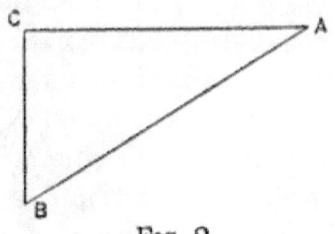

FIG. 2.

by the resistance, and then complete the right-angled triangle. For if ABC is such a triangle,—AB, BC, AC representing respectively the preliminary resultant, the electromotive force of self-induction, and the final resultant at the maximum values,—it is clear that the maximum rate of increase of the resultant electromotive force will be AC × angular velocity. Divide this by the resistance, and the maximum rate of increase in the *current* is obtained, which, multiplied by the coefficient of self-induction, must give the maximum electromotive force of self-induction, from the fundamental conception of that magnitude.

Hence, in symbols, if $r =$ the resistance,
$L =$ the coefficient of self-induction,
$\omega =$ the angular velocity,

$$BC = \frac{AC}{r} \omega L,$$

or $\quad \dfrac{BC}{AC} = \tan BAC = \dfrac{\omega L}{r}.$

If 2T is the period, $\omega = \dfrac{2\pi}{2T}, \therefore \tan BAC = \dfrac{L\pi}{Tr}.$

And since the electromotive force of self-induction must be greatest and $+$ve when the current is changing *through zero from* $+ve$ *to* $-ve$, it is clear that the phases of the electromotive force of self-induction must *follow* those of the final resultant electromotive force at an interval of time represented by a quarter of the period. Thus the above construction is justified.

As many of the problems involving alternating currents can be very well exhibited and solved by geometrical methods, I shall give one or two geometrical propositions which will render the diagrams that may occasionally be required easier of comprehension.

Geometrical Proposition I.

If AB, AC, are two lines in the plane of AX, AY, which revolve round a line drawn through A at right angles to this plane, at a uniform rate, the angle CAB being therefore maintained constant, to find a geometrical expression for the mean value of the product of the projections of AB, AC upon AX.

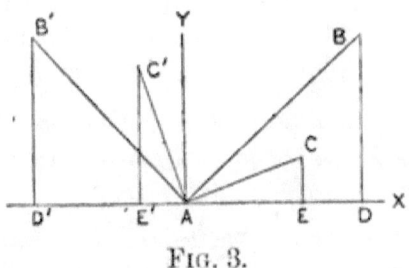

Fig. 3.

From B and C draw BD, CE, perpendicular to AX.

Draw AB′ perpendicular to AB, and make AB′ equal to AB.

Draw AC′ perpendicular to AC, and make AC′ equal to AC. Then AB′, AC′ represent the positions of AB, AC after revolving through one right angle.

Draw B′D′, C′E′ perpendicular to XA produced.

Then the angle AB′D′ is equal to the angle BAD, and the angle AC′E′ is equal to the angle CAE.

Now

$$AE, AD = AC, AB, \cos CAE, \cos BAD,$$

and

$$AE', AD' = AC', AB', \sin AC'E', \sin AB'D$$
$$= AC, AB, \sin CAE, \sin BAD;$$

therefore

$$\frac{AE, AD + AE', AD'}{2} = \frac{AC, AB}{2} \{ \cos CAE \cos BAD + \sin CAE, \sin BAD \}$$

$$= \frac{AC, AB}{2} \{ \cos \overline{CAE - BAD} \}$$

$$= \frac{AC, AB}{2} \cos BAC.$$

But the quantity upon the left of this equation represents the mean value of the product of the projection of AB, AC, upon AX, for two positions of the moving system differing by one right angle. And all the positions of the system may be taken in pairs differing by one right angle. But when two such positions are taken, the mean value is shown by the above equation to be independent of the actual position of the system.

Therefore, the mean value obtained for two such positions is the mean value for all positions, and is given by the above equation, the right hand side of which shows it to be half the product of AB and AC multiplied by the cosine of the angle between them.

It is unnecessary that the lines AB, AC should have a common point. If AB, CD are any two lines situated in one plane, and inclined to one another at a constant angle, while they revolve round any axis at right angles to the plane containing them, the mean value of the product of their projections on any fixed line in the plane containing them is half the product of their lengths multiplied by the cosine of the angle between them.

The application of this proposition to problems in alternating currents is extremely simple.

The power working at any instant in a source of electromotive force is the value of the product of the instantaneous electromotive force in question, and of the instantaneous current.

Suppose one of the lines we have been considering represents the maximum value of an alternating electromotive force varying harmonically, and acting upon a circuit conveying an alternating current also varying harmonically, and that the two have the same period. Suppose the second of the two lines to represent the effective electromotive force through the circuit at a maximum. Then the projections of these two lines may be taken to represent the real value of these two electromotive forces at some instant. If we divide the effective electromotive force by the resistance, we get the real value of the current at the instant, and the product of the electromotive force represented by the projection of the first line into the current at the instant represents the power at the instant exerted by the source of the electromotive force represented by the first line.

Hence, if in the figure AB represent the maximum value of some electromotive force acting upon a circuit whose resistance is R, and AC represent the maximum of the effective electromotive force producing current, then the mean power exerted by the source of electromotive force, whose maximum is AB, is equal to $\dfrac{AB \cdot AC}{2R} \cos BAC$.

Now, suppose that the only two electromotive forces acting upon the resistance R are represented by AB,

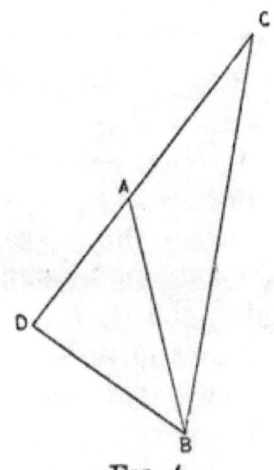

Fig. 4.

BC, when at their maximum, and by the projection of these lines upon any fixed straight line at any other moment, then AC is the effective electromotive force acting upon the resistance R. Join AC, and produce it towards A to D, and draw BD perpendicular to it.

Then, by what has been said, the power exerted by the source of BC is $\dfrac{BC, AC}{2R}$ cos ACB, and the power exerted by the source of AB is $\dfrac{AB, AC}{2R}$ cos CAB.

This in the figure as drawn has a negative value, equal numerically to $\dfrac{AB, AC}{2R}$ cos DAB, which, therefore, represents the power doing work upon the source of AB, and, following the same rule, the effective power sending the current through the circuit is—

$$\dfrac{AC \times AC}{2R} \cos 0 = \dfrac{AC^2}{2R}$$

The three powers of the primary source, of the recipient source, and that heating the circuit, are proportional respectively to

BC cos ACB : AB cos DAB : AC, *i.e.*, to CD : DA : AC.

The electromotive force of self-induction is one which depends for its value upon the rate of increase or decrease of the current. It is greatest when the current is increasing most rapidly, *i.e.*, when the current is zero, and it is least when the current is unchanging, *i.e.*, when it is at a maximum or minimum. It varies, therefore, harmonically, and is, in fact, an electromotive force which must be drawn in such a diagram of electromotive forces as has been considered at right angles to the effective electromotive force. Its mean power, therefore, in a period must be zero.

To find its maximum value, consider that if $2T$ is the complete period, $\frac{2\pi}{2T}$ is the angular velocity of any point in the diagram. And if the diagram revolve round one extremity of the line representing the effective electromotive force, the greatest rate of increase of the effective electromotive force is represented by the velocity of the other point.

If e be the maximum value of the effective electromotive force, $e\frac{2\pi}{2T}$ will represent its greatest rate of increase, and this divided by the resistance R, will represent the greatest rate of increase of the current. If, therefore, the circuit has a self-induction coefficient of the value L, the maximum value of the electromotive force of self-induction will be $L\frac{e}{R}\frac{2\pi}{2T} = e\frac{L\pi}{RT}$; therefore, in the diagram $\frac{L\pi}{RT}$ must be equal to the tangent of the angle between the effective electromotive force and the resultant of the impressed electromotive forces.

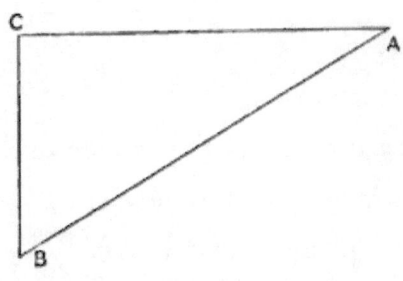

FIG. 5.

Thus, in the figure, if AB represent the resultant of the impressed electromotive forces, and AC is drawn so that tangent $CAB = \frac{L\pi}{RT}$, and BC is drawn perpendicular to AC, then BC represents the electromotive force of self-induction and AC the effective electromotive force.

Suppose, now, that AB, BC are the revolving representatives of two electromotive forces. Then AC is their resultant; CAE is an angle whose tangent is equal to $\frac{L\pi}{Tr}$, as explained; CE, BF are perpendiculars upon AE. Then AE is the final resultant or effective electromotive force, merely requiring division by the resistance to give the current.

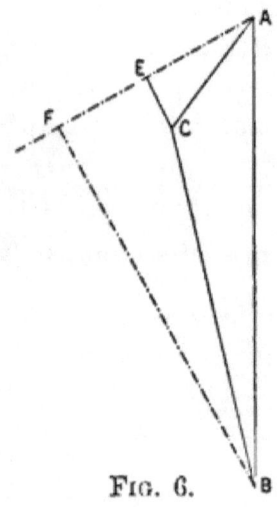

FIG. 6.

The power derived from the source of AB is $AF \frac{AE}{2r}$; the power transferred to the source of BC is $FE \frac{AE}{2r}$; and the power heating the circuit is $AE \frac{AE}{2r} = \frac{AE^2}{2r}$.

As regards the projection of BC, viz. FE, since (as here drawn) FE is in a contrary direction to AE, there is a transfer of power to its source. Had F been situated nearer to A than E is, the source of BC would do work and assist in heating the circuit. This obviously depends upon whether BC, AE are inclined to one another at an angle greater or less than a right angle.

If we denominate these three powers as the power of the active source, the power of the recipient source, and the heating-power, they will be to each other in the proportion

$$AF : FE : AE;$$

and the efficiency of transmission will be $\frac{FE}{AF}$, the ratio of waste being $\frac{AE}{AF}$.

The electromotive force of self-induction, properly speaking, exerts no power in a period. It happens, however, that the cores of electro-magnets become warm when subjected to an alternating current; and this effect is generally attributed to the heat from the circuit reaching them by conduction. Some high authorities have held that the power due to any in-

FIG. 7.

duced electromotive force must be zero, and that the change of electro-kinetic momentum involves no loss of energy. It is certain that if there be any electromotive force doing work, its projection on the line of effective electromotive force cannot vanish; and if any work is done upon the cores by the changes of polarity in one period, the effective electromotive force will be diminished by the amount of one opposite to it in phase and of such a value that if multiplied by the reduced effective one and divided by twice the resistance, the result will be the power producing the work observed to be done upon the cores. Such an electromotive force is an induced one; but it differs from that of self-induction, as usually understood, in having its phase in exact opposition to that of the current. In such a case the diagram of electromotive forces would be as shown above (Fig. 7).

Here AB is the impressed electromotive force.

AD is the effective electromotive force.

CD is the ordinary electromotive force of self-induction.

BC is the electromotive force resulting from magnetic viscosity, and DC, DA are connected by the relation,—

$$\frac{DC}{DA} = \frac{L\pi}{TR}$$

The result of a large magnetic viscosity would be, therefore, to diminish the effective electromotive force in comparison with the impressed one, but at the same time to bring the two more nearly into the same phase; whereas the effect of an increase in the co-efficient of self-induction, as commonly understood, is to diminish the effective electromotive force in comparison with the impressed one, and at the same time to increase the difference of phase.

Most problems relating to self and mutual induction, and to the action of condensers at particular points, may be treated geometrically in the way indicated, as will be exemplified occasionally in the following chapters, though the analytical method is probably better in cases of distributed capacity.

CHAPTER II.

MUTUAL-INDUCTION.

For the sake of brevity it will be well to describe a magnitude undergoing harmonic variation by its maximum value. Thus, suppose, in Fig. 8, BD, the projection of a constant line, AB, upon the axis of Y, to represent the electromotive force at some instant acting through a circuit, and that the electromotive force undergoes the variations which this projection undergoes, it will be convenient to describe the electromotive force by its maximum value AB.

And similarly with currents. If a current is represented always by the projection of the magnitude representing its maximum, we may talk of the maximum value as the current.

Language based upon this convenient mode of expression has, in fact, been used towards the close of the previous article. We may employ the first proposition there given to the determination of the reading of a dynamometer subjected to alternating currents.

Suppose, in Fig. 1, that AB represents the current through one of the coils of the dynamometer, and that AC represents the current through the other coil. Then the average value of the product of these currents will be $\dfrac{AC\,AB}{2}\cos BAC$.

But it is the average value of the product of these currents which will be proportional to the reading of the instrument.

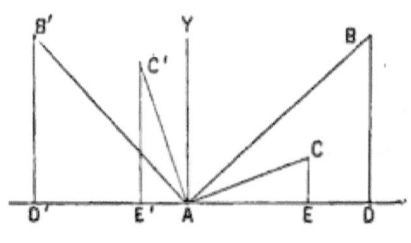

Fig. 8.

Moreover, in the ordinary case the currents in the two coils are identical in value and coincident in phase, the coils being in series.

The first consideration requires AB to be equal to AC, and we may take c as the value of each of them. The second consideration requires cos BAC to be equal to unity, since BAC $= 0$.

Therefore, the reading of the dynamometer will be proportional to $\dfrac{c^2}{2}$.

Thus the reading of the instrument which is produced by an alternating current c (*i.e.*, whose maximum is c) will be produced by a uniform current whose value is $\dfrac{c}{\sqrt{2}}$ or $\cdot 707\, c$.

The method of compounding the electromotive forces acting in one circuit and having the same period, but differing in phase, after the manner of forces and velocities in mechanics, is evidently true when we consider that the real electromotive force is the sum of the projections of each, individually projected, upon one straight line.

The graphic diagram resulting from the application

of this method may be advantageously employed in exhibiting the effects of an alternating E.M.F. acting on a circuit possessing, besides self-induction, mutual induction on and from another circuit possessing self-induction in its turn.

As self-induction is defined to be the electromotive force acting through a circuit due to the increase or decrease of the current in the circuit itself, so mutual induction is the electromotive force produced in one circuit by the changing current in a neighbouring circuit.

In each case the rate of change in the current requires to be multiplied by a coefficient to give the induced electromotive force.

L is the usual symbol for the coefficient of self-induction, and M is the coefficient of mutual induction.

Hence, if we multiply the instantaneous rate of increase of the current in the first of two coils by M, the coefficient of mutual induction between that coil and the second, we obtain the instantaneous electromotive force due to this increase in the second coil.

Let L be the self-induction coefficient of the primary circuit.

Let L' be the self-induction coefficient of the secondary circuit.

Let R and r be the resistances respectively of these two circuits, and

M the coefficient of mutual induction between them.

Take a line CF of magnitude h and draw CE at right angles to CF, and in magnitude equal to $\frac{M\pi}{RT}h$, so that, in fact, $\frac{M\pi}{RT} = \tan$ CFE. On CE as diameter describe a semi-circle on the side remote from CF, and set off the angle ECA, whose tangent is equal to $\frac{L'\pi}{rT}$. From F draw FB perpendicular to AC produced, and produce BF towards F to G, so that $FG = \frac{M\pi}{rT}AC$. In CE take CD, so that CD bears to CE the same proportion which L bears to M. Join DG, and call it e.

Then if e is the impressed electromotive force in the primary circuit, in phase and magnitude, CF or h will be the effective electromotive force in the primary

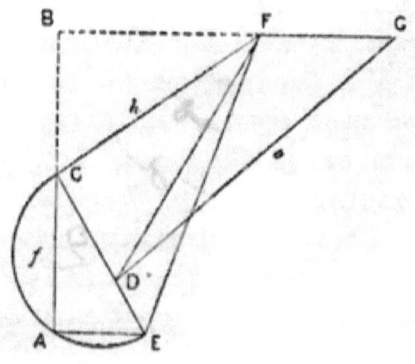

Fig. 9.

circuit, and CA, which call f, the effective electromotive force in the secondary circuit, in phase and magnitude. In other words, dynamometers inserted in these two circuits will give the same reading as if the *constant* E.M.F.'s $\frac{h}{\sqrt{2}}$ and $\frac{f}{\sqrt{2}}$ were inserted in the circuits respectively, the angles of torsion being proportional to the squares of these lines CF, AC respectively.

If we write
$$\tan a = \frac{L\pi}{RT},$$
$$\tan \beta = \frac{L'\pi}{rT},$$
$$\tan \gamma = \frac{M\pi}{rT},$$
$$\tan \delta = \frac{M\pi}{RT},$$
then $CFD = a$, $ACE = \beta$, $CFE = \delta$, and from the geometry of the figure we have
$$\left. \begin{array}{l} f = h \cos \beta \tan \delta \\ e^2 = f^2 \tan^2 \gamma + \dfrac{h^2}{\cos^2 a} + 2fh \dfrac{\tan \gamma \cos \overline{\beta + a}}{\cos a} \end{array} \right\}$$
Therefore we have for the equation connecting e and h
$$e^2 = \frac{h^2}{\cos^2 a} \{1 + \cos^2 a \cos^2 \beta \tan^2 \gamma \tan^2 \delta + 2 \cos a \cos \beta \tan \gamma \tan \delta \cos \overline{a + \beta} \},$$
and for the equation connecting e and f
$$e^2 = \frac{f^2}{\cos^2 a \cos^2 \beta \tan^2 \delta} \{1 + \cos^2 a \cos^2 \beta \tan^2 \gamma \tan^2 \delta + 2 \cos a \cos \beta \tan \gamma \tan \delta \cos \overline{a + \beta} \}$$
If we write $\tan \phi$ for $\cos a \cos \beta \tan \gamma \tan \delta$, for brevity's sake, these equations become:—
$$e^2 = \frac{h^2}{\cos^2 a \cos^2 \phi} \{1 + \sin 2\phi \cos \overline{a + \beta}\},$$
$$e^2 = \frac{f^2 \tan^2 \gamma}{\sin^2 \phi} \{1 + \sin 2\phi \cos \overline{a + \beta}\};$$
or, to express the effective electromotive forces separately in terms of e,
$$h = \frac{e \cos a \cos \phi}{\sqrt{1 + \sin 2\phi \cos \overline{a + \beta}}},$$
$$f = \frac{e \sin \phi \cot \gamma}{\sqrt{1 + \sin 2\phi \cos \overline{a + \beta}}},$$

whence the expressions for the currents $\dfrac{h}{R}$ and $\dfrac{f}{r}$, in terms of the coefficients of self and mutual induction, the resistances and the period, become, by the proper substitutions

$$\dfrac{h}{R} = \dfrac{eT\,(r^2T^2 + L'^2\pi^2)^{\frac{1}{2}}}{\{(M^2\pi^2 + RrT^2 - LL'\pi^2)^2 + \pi^2\,T^2\,(Lr + L'R)^2\}^{\frac{1}{2}}}. \quad (a)$$

$$\dfrac{f}{r} = \dfrac{eTM\pi}{\{(M^2\pi^2 + RrT^2 - LL'\pi^2)^2 + \pi^2\,T^2\,(Lr + L'R)^2\}^{\frac{1}{2}}}. \quad (\beta)$$

The relation which the current in the secondary coil bears to that in the primary is simply

$$\dfrac{M\pi}{(r^2T^2 + L'^2\pi^2)^{\frac{1}{2}}} \quad\cdot\quad\cdot\quad\cdot\quad\cdot\quad (\gamma)$$

and the relation of the dynamometer readings in the two circuits will be as the square of this expression.

The interesting experiments of Mr. Willoughby Smith on the effects produced by inserting masses, or, as they were described, screens of metal in the field of two coils on their mutual induction, illustrate this proposition. Though the undulations of the primary electromotive force were not in this case harmonic, but were produced by mere commutation at the poles of a battery, yet there is no doubt that the effects produced in the secondary coils by this mode of generation would approximate to those illustrated here. From the figure it is clear that a large increase in the value of M will greatly diminish the proportion which h bears to e, but unless L' is largely increased, the increase of M will not so largely alter the value of CE, upon which, with L', f depends.

In the lecture which the writer had the good fortune to hear from Mr. Willoughby Smith, the chief omission seemed to be that of a dynamometer in the primary coil. It is highly probable that a great falling off would have been noticed in the reading of such an instrument, when the metal was placed in the field, even when the dynamometer in the secondary circuit showed no falling off of current. The faulty use of the idea of *screening* would have been made very manifest by such an experiment, for it would have seemed as if the screen had been enabled to *maintain* the effect of the *radiation* on the recipient circuit when the source of the radiation itself had become weaker. I have, in fact, illustrated this point in an experiment.

A battery and electric tuning fork were joined in circuit with the primary coil of a Ruhmkorff apparatus of which the usual vibrator was constrained in position to maintain contact, the tuning fork taking its place as vibrator. In the secondary circuit was a Bell telephone. The core was removed and the apparatus set to work, both the tuning fork and telephone singing in unison. On pushing the core into its place in the centre of the coils, the intensity of the tone of the telephone was increased, but that of the tuning fork was diminished. The period was $\frac{1}{450}$ of one second, the tuning fork giving the A above tenor C. Probably a lower note would have increased the effect observed. To render the effect on the tuning fork more evident, it is well to silence the telephone. This can be conveniently done by holding it mouth downwards, and burning a whisp of paper about 6 in. beneath.

On the other hand, the figure and the formulæ show that when the period is diminished, other things remaining the same, both the effective electromotive force in the primary and that in the secondary circuit grow smaller, compared with the impressed electromotive force. But the equation (γ) indicates that when T is diminished, *i.e.*, when the alternations become more rapid, the relation of the current in the secondary circuit to that in the primary circuit becomes greater; thus, in that part of Mr. Smith's experiments which dealt with an augmentation of the rate of alternation, the energy could hardly be said to be *screened* when a dynamometer in the primary coil would have demonstrated a greater falling off in the current in that circuit than occurs in the secondary circuit. To describe such an effect as due to screening would be as if one suffering from the radiation from a grate should pull between himself and it a screen, which, by being drawn into the middle position, should, by suitable mechanism, cause a bucket of water to be emptied into the fire, and were then to declare that the screen had stopped the radiation by virtue of being a screen.

CHAPTER III.

CONDENSERS.

In applying the geometrical method to the representation of the effects upon the carrying power of a circuit of electric capacity it will be necessary to bear in mind carefully what is meant by capacity. An arrangement is said to possess capacity when the arrival and accumulation of a certain finite quantity of electricity is necessary before it can exhibit a rise in potential. If the rise in the potential is proportional to the net arrival of electricity, the capacity of the condenser is said to be constant, and is measured by that amount of electricity necessary for a rise of potential of a unit amount. It has been denied that a voltameter is a condenser, but without justice. The arrival and accumulation of electricity is necessary before the electrodes of the voltameter exhibit difference of potential. When this takes place, it is accompanied, not perhaps by what is sometimes understood as a static charge, but by a decomposition of the fluid, and an annexation, incipient or otherwise, by the electrodes of the products of the electrolysis. In an incipient condition, with which in the question of alternations we should have to deal, it is not certain how the quantity of electricity which has reached the voltameter is connected with the difference of potential exhibited by the electrodes.

That is to say, the condenser of this form may not have a constant capacity, and therefore might disturb the harmonic nature of the alternations, even when the variation of the impressed electromotive force was truly harmonic. But with the more usual form of condenser, where there exists a constant capacity, there will be, indeed, upon its introduction into the system, a re-arrangement of the values of the current in different parts, but each will remain harmonic in character.

Suppose the straight lines AB, BC, CD, etc., being the consecutive sides of an unclosed polygon, represent the values of *currents* conveying electricity harmonically to and from a point, with the same period of variation.

Then, since the rate of arrival of electricity is represented at any instant by the sum of the projections of the individual lines upon some fixed line, and the sum of such projections is equal to the projection of the line joining AD, the extreme points, therefore the arrival of electricity will be that due to *one* current of the value AD.

Thus, however many harmonic currents of the same period are flowing into and out of a condenser, and whatever their differences of phase, their effect in charging that condenser will be that due to *one harmonic current* of the same period, and in a certain determinate phase as regards the actual currents. This current may be called the effective current.

It will be necessary to consider, therefore, the effect upon a condenser of constant capacity of the arrival of electricity at a rate undergoing harmonic variation, for that rate is the effective current.

While the current is positive, the charge is always increasing; and, *vice versâ*, the charge of the condenser is always decreasing when the current flows away from

CONDENSERS. 25

the condenser. Hence the charge will be greatest when the current is changing from positive to negative through the value zero, and least when the current is changing from negative to positive through the value zero.

Let us take this latter point to start from, and consider what the accumulation of electricity will be in a given time t. If T is half the period, the angle through which the line representing the current will revolve in time t is equal $t\frac{\pi}{T}$. Call this angle a. Let θ represent any angle which the line has traversed between zero and a by a given point of time.

Then, if c is the maximum value of the current, $c \sin \theta$ will be its value at the particular point of time considered. Now, supposing we knew the average value of this current between the values of θ, zero and a, we should only have to multiply this average into the time t to find the value of the electrical accumulation. The factor c is constant, and therefore the accumulation depends upon the average value of $\sin \theta$ between 0 and a.

The following subsidiary proposition is a mode of finding the value required without the use of the integral calculus.

GEOMETRICAL PROPOSITION II.

To find the average value of $\sin \theta$ and $\cos \theta$ when θ varies from 0 to a uniformly.

Let S(a), C(a), represent respectively the average values of $\sin \theta$ and $\cos \theta$, when θ varies uniformly from 0 to a.

Now, the cosine of any angle between 0 and a may

be expressed by $\cos\left(\frac{a}{2}+\phi\right)$ or $\cos\left(\frac{a}{2}-\phi\right)$; where ϕ may have all values from 0 to $\frac{a}{2}$.

Therefore all the values of θ may be grouped into a number of pairs, represented by $\left(\frac{a}{2}+\phi\right)$ and $\left(\frac{a}{2}-\phi\right)$ where ϕ varies from 0 to $\frac{a}{2}$.

Now, $\cos\left(\frac{a}{2}+\phi\right) = \cos\frac{a}{2}\cos\phi - \sin\frac{a}{2}\sin\phi$,

and $\cos\left(\frac{a}{2}-\phi\right) = \cos\frac{a}{2}\cos\phi + \sin\frac{a}{2}\sin\phi$.

Therefore, the average value of the pair of cosines, $\cos\left(\frac{a}{2}+\phi\right)$ and $\cos\left(\frac{a}{2}-\phi\right)$ is $\cos\frac{a}{2}\cos\phi$.

Thus the average value of $\cos\theta$, where θ varies from 0 to a uniformly, depends upon the average value of $\cos\phi$, where ϕ varies uniformly from 0 to $\frac{a}{2}$, as in the following expression:—

$$C(a) = \cos\frac{a}{2} C\left(\frac{a}{2}\right).$$

Similarly $\quad C\left(\frac{a}{2}\right) = \cos\frac{a}{4} C\left(\frac{a}{4}\right).$

Therefore $C(a) = \cos\frac{a}{2}\cos\frac{a}{4} \cdot \cdot \cdot \cdot \cos\frac{a}{2^n} C\left(\frac{a}{2^n}\right).$

The angle $\frac{a}{2^n}$ becomes indefinitely small when n is made infinitely large, and evidently the average value of the cosine of a very small angle approaches unity, therefore $C\frac{a}{2^n}$ is equal to unity when n is made infinitely great.

Therefore $C(a) = \cos\frac{a}{2} \cos\frac{a}{4} \cos\frac{a}{8} \ldots$ ad infinitum.

But by ordinary trigonometry

$$\sin a = 2 \cos\frac{a}{2} \cdot \sin\frac{a}{2}$$

$$= 2^2 \cos\frac{a}{2} \cdot \cos\frac{a}{4} \cdot \sin\frac{a}{4}$$

$$= a \cdot \frac{2^n}{a} \cdot \cos\frac{a}{2} \cos\frac{a}{4} \cos\frac{a}{8} \ldots \cos\frac{a}{2^n} \sin\frac{a}{2^n}$$

$$= a \cdot \cos\frac{a}{2} \cos\frac{a}{4} \cos\frac{a}{8} \ldots \cos\frac{a}{2^n} \left(\frac{\sin\frac{a}{2^n}}{\frac{a}{2^n}} \right)$$

and when n is infinitely larger,

$$\frac{\sin\frac{a}{2^n}}{\frac{a}{2^n}} = 1.$$

Therefore, $\sin a = a \cos\frac{a}{2} \cos\frac{a}{4} \cos\frac{a}{8} \ldots$ ad infinitum, and hence

$$C(a) = \frac{\sin a}{a} \quad \ldots \quad (a)$$

or the average value of $\cos\theta$, when θ varies uniformly between 0 and a, is $\frac{\sin a}{a}$.

Similarly,

$$\sin\left(\frac{a}{2} + \phi\right) = \sin\frac{a}{2} \cos\phi + \cos\frac{a}{2} \sin\phi,$$

and $\sin\left(\frac{a}{2} - \phi\right) = \sin\frac{a}{2} \cos\phi - \cos\frac{a}{2} \sin\phi.$

Therefore, the average value of the pair of sines, $\sin\left(\frac{a}{2} + \phi\right)$ and $\sin\left(\frac{a}{2} - \phi\right)$ is $\sin\frac{a}{2} \cos\phi$.

Thus the average value of $\sin \theta$, when θ varies from 0 to a uniformly, depends upon the average value $\cos \phi$, where ϕ varies from 0 to $\frac{a}{2}$, as in the following expression:—

$$S(a) = \sin \frac{a}{2} C\left(\frac{a}{2}\right).$$

But by the former portion of the proposition the value of $C\left(\frac{a}{2}\right)$ has already been determined (a) to be

$$\frac{\sin \frac{a}{2}}{\frac{a}{2}}$$

Therefore $S(a) = \dfrac{\sin^2 \frac{a}{2}}{\frac{a}{2}} = \dfrac{1-\cos a}{a}$; . . . . ($\beta$)

that is to say, the average value of $\sin \theta$, when θ varies from 0 to a, is

$$\frac{1-\cos a}{a}.$$

We can now deal with the electric problem, for the supply of electricity given by the harmonic current in a time t dating from the zero value of the current will be $tc S\left(\dfrac{t\pi}{T}\right)$

or $tc \times \dfrac{1-\cos \frac{t\pi}{T}}{\frac{t\pi}{T}} = \dfrac{Tc}{\pi}\left(1-\cos \frac{t\pi}{T}\right) = \dfrac{Tc}{\pi} - \dfrac{Tc}{\pi}\sin\left(\dfrac{\pi}{2} - \dfrac{t\pi}{T}\right)$

. . . . (γ)

and the total supply during one half period is therefore equal to $\dfrac{2Tc}{\pi}$, which is the value of the above expression when t is equal to T. During the second half period this quantity of electricity is abstracted.

During the first quarter period the supply will be $\frac{Tc}{\pi}$, or half the entire quantity, and as it is convenient always to measure from a mean value of any harmonically varying magnitude, we may consider $\frac{Tc}{\pi}$ to be the addition or withdrawal of electricity in any of the quarter periods, to or from the mean value.

The expression (γ) shows that the variation in the supply is harmonic, and differs by a quarter period in phase from that of the current in the direction of retardation; *i.e.*, it does not come up to its maximum until the current has passed its maximum, the interval between the two events being a quarter of a period, or $\frac{T}{2}$.

Hence, if the condenser be one whose capacity is constant, the potential difference of its plates will undergo harmonic variation, and in a diagram of electromotive forces will be drawn at right angles to any line which is in the same phase as the effective current into the condenser, in the direction of retardation. If C is the capacity of the condenser, the amplitude of the difference of potential of its plates will be $\frac{1}{C}\frac{Tc}{\pi}$ which E.M.F. will occur when the effective current is changing its direction.

We shall next consider the effect of connecting two points of a circuit with the plates of a condenser.

CHAPTER IV.

CONDENSER IN CIRCUIT.

BEFORE discussing some of the effects upon the practical applications of electricity of coefficients of mutual and self-induction, it will be well to consider the action of a condenser whose plates are joined to different points of a circuit conveying an alternating current, for we shall find that in cases where both capacity and coefficients of induction exist, they are capable not only of modifying the effects of each other, but of masking the effects of resistance. It may even be possible, by suitable adjustment of the quantities concerned, to put through a resistance which is not closed a current of the same amount which would flow through it if it were closed, but denuded of the condensing apparatus and induction coefficient.

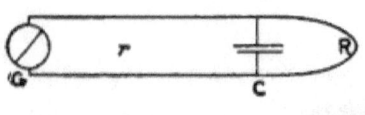

FIG. 10.

In Fig. 10 suppose that G is a source of harmonically varying electromotive force, working into a circuit whose total resistance is $\overline{R+r}$, and that at C there is a condenser of capacity C connected with the circuit in such a way that the section of the circuit terminated by the plates of the condenser and containing the source G has a resistance r, and that the remote section has a resistance R. We shall also suppose that there is no self-induction in either section of the conductor. Take a straight line, O C, on any convenient scale to represent the electromotive force of the source, and divide OC in D so that OD : DC :: r : R.

Calculate the quantity $\dfrac{C\pi R r}{T(R+r)}$, where T is half the period. From the nature of the quantities involved this quantity is necessarily positive and numerical, and it may therefore be represented by the tangent of an angle in a diagram where the lines, as in the case before us, are electromotive forces.

Set off the angle OCE of such a value that its tangent has this value, and in the direction of retardation. From D draw DE perpendicular to CE, and join EO. Then OE and EC will represent in phase and magnitude the effective electromotive forces in the two sections. That is to say, the current generated in r will be that due to an E.M.F. of the value OE operating upon a simple circuit of resistance r, and the current generated in R will be that due to an E.M.F. of the value EC operating upon a simple circuit of resistance R. To prove this, produce CE to F, and draw OF perpendicular to CF.

In the first place it is to be observed that the difference of potential of the plates of the condenser constitutes the effective electromotive force in the circuit R. Also the difference between the generated E.M.F. and that due to the plates of the condenser constitutes the effective E.M.F. in the circuit r.

But the generated E.M.F. is OC. If, therefore, EC does represent the E.M.F. due to the condenser—*i.e.*, the effective E.M.F. in section R—it follows that OE will represent the effective E.M.F. in section r. Thus the two points to be proved stand or fall together.

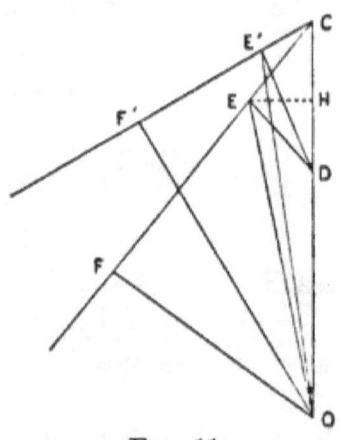

Fig. 11.

If EC is not the effective E.M.F. in circuit R, let E'C be the value in phase and magnitude. Join E'O, E'D, and draw OF' parallel to DE'.

Now, because E'C represents the effective E.M.F. in R, therefore OE' represents the effective E.M.F. in r.

Therefore $\dfrac{E'C}{R}$ is the effective current in R,

and $\dfrac{OE'}{r}$ is the effective current in r,

but F'E' : E'C :: OD : DC, because OF' is parallel to DE'.

Therefore F'E' : E'C :: r : R,

and hence $\dfrac{F'E'}{r} = \dfrac{E'C}{R}$.

Therefore $\dfrac{F'E'}{r}$ is the effective current in R,

and since $\dfrac{OE'}{r}$ is the effective current in r,

therefore $\dfrac{OE'}{r}$ and $\dfrac{E'F'}{r}$ are those currents considered as

CONDENSER IN CIRCUIT.

conveying electricity *into* the condenser, the positive current in R being of course a negative current as regards charging the condenser. But these are the only two currents affecting the condenser; therefore $\frac{OF'}{r}$ is the effective charging current of the condenser—*i.e.*, the phase and magnitude of the effective E.M.F. through a resistance r, and producing the effective charging current, is represented by the line OF'. But, by what was proved in the last chapter, the phase of the charging current must differ by a quarter period from that of the difference of potential of the plates of the condenser.

Therefore must OF' be at right angles to CE', and the angle DE'C be a right angle. Therefore E' must lie in the circumference of the semi-circle described upon CD as diameter.

Further, since $\frac{OF'}{r}$ is the effective charging current, the maximum accumulation of electricity is $\frac{OF'}{r}\frac{T}{\pi}$, and therefore, if C is the capacity of the condenser, the difference of potential of the plates of the condenser will be $\frac{OF'}{r}\frac{T}{\pi}\frac{1}{C}$, which therefore should be equal to E'C.

But OF' : E'D :: CO : CD by similar triangles.

$$:: \overline{R+r} : R$$

$$\therefore OF' = E'D\,\frac{\overline{R+r}}{R}$$

Hence $\quad E'D\,\dfrac{\overline{R+r}\,T}{r\,R\,\pi\,C} = E'C$

Therefore $\quad \dfrac{E'D}{E'C} = \dfrac{C\pi R r}{T.R+r}$

or tan DCE' = tan DCE.

Al. Cu.

Therefore CE′ must coincide with CE in direction, and because E′ lies upon the semi-circle on DC, it can occupy no other position than E; or EC is the effective E.M.F. in R, and by consequence OE is the effective E.M.F. in r.

The only point remaining to be justified is the direction in which the angle OCE has been taken. Had it been taken on the other side of OC, it is clear that EC would be at a maximum a quarter period *before* $\frac{OF}{r}$, the effective charging current, which would be absurd. Hence the construction is completely justified.

Now, it is evident, from the mode of construction, that OE must always be greater than OD. But had there been no condenser in the system, it is clear that OD would have been the effective E.M.F. through the resistance r. The effect of the condenser, then, has been to increase the effective E.M.F. in this section, and of course also to increase the current resulting from it.

On the other hand, EC must necessarily be less than CD, which is the longest side of a right-angled triangle, of which EC is one of the other sides. But without the condenser CD would have been the effective E.M.F., through the section whose resistance is R. The effect of the condenser in the system upon this portion of it will be, therefore, to diminish the current.

Hence, if the whole circuit before the introduction of the condenser contains a number of incandescent lamps in series, the joining up with the plates of the condenser will have opposite effects upon the lamps in the two portions of the circuit. The light of those situated between the condenser and the generator will be augmented, and that of those in the remote section will grow less intense. This effect depends entirely upon the value of the tangent of the angle OCE, which is equal to $\dfrac{C\pi Rr}{T(R+r)}$, and therefore increases as the capacity of the condenser increases, and as the period diminishes.

A further question is that of the change of power employed. By what was demonstrated in the first article, the power of the source of any harmonic source of E.M.F. is equal to half the product of the E.M.F. and the current passing through it, multiplied by the cosine of their difference of phase.

In the case before us, the E.M.F. of the generator is OC; the current through it is $\dfrac{OE}{r}$.

Therefore the power is $\dfrac{OC \cdot OE}{2r} \cdot \cos COE$.

If EH is drawn perpendicular to OC,
$$OH = OE \cos \cdot COE.$$

Therefore the power employed is $\dfrac{OC \cdot OH}{2r}$.

But had there been no condenser, or if the capacity of the condenser vanishes, OH would have been coincident with OD, and the power would have been $\dfrac{OC \cdot OD}{2r}$.

Therefore the effect of the condenser has been to demand a change of power in the ratio of OD : OH.

Since OH must always be greater than OD, the change is always in the direction of augmentation, and we are therefore justified in thinking that such an arrangement might be beneficially employed in regulating the intensity of lamps; for though in the remote section the light suffers by the introduction of the condenser, there is upon the whole an augmentation of power at work.

The actual power in the two sections of the circuit will be $\frac{OE^2}{2r}$ in the generator's section, and $\frac{EC^2}{2R}$ in the remote section.

These, of course, when added together, are equal to the power of the source, *i.e.*, to $\frac{OC.OH}{2r}$, which any one can verify from the geometry of the figure.

It may be interesting to note, however, that we can exhibit the powers in the two circuits as the products of certain lines in the figure divided by the *same* denominator, so that they can be compared by the mere multiplication of measurements taken with a pair of dividers.

The power in the remote circuit is $\frac{EC^2}{2R}$; but EC : EF :: R : r, therefore $\frac{EC}{R} = \frac{EF}{r}$, and the power becomes $\frac{EC.EF}{2r}$, while that in the generator's circuit is $\frac{OE^2}{2r}$, and the total power is $\frac{OC.OH}{2r}$.

Therefore they can be compared with each other by measuring EC.EF, OE², OC.OH.

Also, since CH : CE :: FE : DO. Therefore, CE.FE = CH.DO.

Therefore, the three powers are again as
$$CH.DO : OE^2 : OC.OH;$$
and we have only to divide by $2r$ to obtain the absolute powers.

CHAPTER V.

SEVERAL CONDENSERS.

WE may easily extend the reasoning of the case when the circuit is bridged by one condenser to that in which at several points condensers are situated, as indicated by Fig. 12.

In this figure G is a generator, and the circuit is bridged at various points C_1 C_2, etc., by condensers of capacity C_1 C_2, etc. R_1 is the resistance of the remotest section, R_2 that of the section between condensers C_1 and C_2, R_3 that of the section between condensers C_2 and C_3, and so on up to C_n, where the last condenser is situated. R is the resistance of that section which contains the generator. Suppose, also, that there is no self-induction in any of the sections.

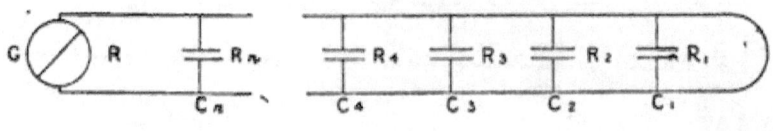

FIG. 12.

So far as the two remotest sections are concerned, the distribution of flow will follow that of the case already given, when the potential difference of the plates of the condenser C_2 takes the place of the impressed E.M.F. Suppose, therefore, $E_1 C$ (Fig. 13) to be the potential difference of the plates of condenser C_1, which for brevity we will call the E.M.F. of condenser C_1.

Set off the angle $E_1 CD_1$ in the direction of advancement so that its tangent is equal to $\dfrac{C_1 \pi R_1 R_2}{T(R_1 + R_2)}$. Draw $E_1 D_1$ perpendicular to $E'C$, cutting CD_1 in D_1, and produce CD_1 to E_2, so that $CD_1 : D_1 E_2 :: R_1 : R_2$. Join $E_1 E_2$. Then $E_2 C$ is the impressed E.M.F., so far as the two remotest sections are concerned. Therefore, $E_2 C$ is the E.M.F. of condenser C_2, and $\dfrac{E_2 E_1}{R_2}$ is the current in R_2.

Now, set off the angle $E_2 CD_2$ in the direction of advancement such that its tangent is equal to $\dfrac{C_2 \pi R_2 R_3}{T(R_2 + R_3)}$, and draw $E_2 D_2$ perpendicular to CE_2. Produce CD_2 to

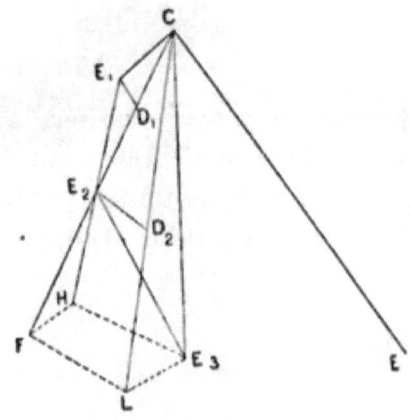

Fig. 13.

L, so that $CD_2 : D_2 L :: R_2 : R_3$, and draw LF parallel to $D_2 E_2$ to cut CE_2 produced in F.

Produce $E_1 E_2$ to H, so that $E_1 E_2 : E_2 H :: R_2 : R_3$. Join

FH, and draw HE_3 parallel to and equal to FL. Join CE_3, $E_3 E_2$. Then, E_3C is the E.M.F. of condenser C_3, and $\frac{E_3 E_2}{R_3}$ is the current in R_3.

These points are proved by the following considerations:—

Firstly, $\frac{LF}{R_3}$ is the effective charging current for the condenser C_2, because $\frac{LF}{FC} = \tan FCL = \frac{C_2 \pi R_2 R_3}{T(R_2 + R_3)}$ by construction, and $\frac{FC}{E_2C} = \frac{R_2 + R_3}{R_2}$.

Therefore, $\frac{LF}{E_2C} = \frac{C_2 \pi R_3}{T}$, or $\frac{LF}{R_3} \frac{T}{\pi} \frac{1}{C_2} = E_2C$.

Therefore, by the formula given at the bottom of page 29, $\frac{LF}{R_3}$ is the effective current charging the condenser C_2, and LF being at right angles to CE_2, is in its right phase.

But E_3H is equal and parallel to LF; therefore, $\frac{E_3H}{R_3}$ is the effective charging current for the condenser C_2.

Secondly, one actual current affecting the condenser C_2 is $\frac{E_2 E_1}{R_2}$, which flows away from the condenser through the section R_2.

Considered as an accumulating current it is $\frac{E_1 E_2}{R_2}$; but $E_2H : E_1 E_2 :: R_3 : R_2$. Therefore, $\frac{E_2 H}{R_3} = \frac{E_1 E_2}{R_2}$.

Thus $\frac{E_2 H}{R_3}$ is one of the currents composing the effective charging current $\frac{E_3 H}{R_3}$. Therefore, $\frac{E_3 E_2}{R_3}$ is the other component. That is to say, that E_3E_2 is the effective E.M.F. in the circuit R_3.

This again being the resultant of the combined simultaneous action upon the resistance R_3 of the E.M. forces of the two condensers C_2 and C_3, and E_2C being that of the condenser C_2, it follows that E_3C is the E.M.F. of the condenser C_3.

It is clear that by extending the construction in the same way until we have taken all the sections into consideration, we shall ultimately arrive at one line, C E, which will represent the E.M.F. of the machine necessary to generate the currents indicated.

It would be very hard to construct a figure from the final line C E; but this process is unnecessary, because the actual value of the impressed E.M.F. would only affect the scale upon which the various lines should be measured, but not the angles or relative sizes of the lines in the diagram. A proper scale being therefore furnished from the final line to represent the value of the E.M.F. of the generator, it may be applied at once to the diagram to indicate—(1) the effective E.M.F. in each section as given by such lines as E_3E_2, E_2E_1, which, being divided by the resistance in the particular section, will give the current—(2) the difference of potential of the plates of the various condensers as given by CE_2, CE_3, etc.

From the diagram it is clear that the general effect of condensers placed locally as bridges in a circuit is twofold. In the first place, when the sections of the circuit are considered in order, there is a continual delay in phase as we recede from the generator. In the second place, the current in the nearer sections is increased, and in the remoter sections diminished by the introduction of the condensers.

If in such a circuit there were a number of incandescent lamps in series, the effect of joining up with the condensers would be, that the lamps nearer the generator would shine more brightly, and those most remote would become duller, while there might be some in the intermediate sections which would indicate no change.

These observations serve to indicate the nature and reason of the decay in an alternating current in a long cable possessing capacity, which will be dealt with more minutely later on. They also explain the error in estimating an electric current by dividing the electromotive force by the resistance, when that electromotive force is a rapidly variable one, and there happens to be any great capacity for electricity involved in the system. It has been already indicated that a voltameter possesses capacity, and there is strong reason to believe that the tissues of the human body possess considerable capacity, probably in the same way as a voltameter possesses it.

If two Geissler tubes are placed in series in the secondary circuit of a Ruhmkorff apparatus, and the pole of a condenser is connected with their point of junction, then either tube may have the intensity of its light increased by the junction of the other pole of the condenser with that electrode of the secondary coil to which the other tube is attached. That other tube suffers a diminution or extinction of its light at the same time. It is in what has been called the remote section. A precisely similar effect is the result of using the two hands as the poles of the condenser. Of course this experiment lies open to the suggestion that it is the conductivity of the body which, when inserted in the system in the above manner, would relieve the remote section, and throw more current through the near one, that produces the effect observed. But there are other facts pointing to the conclusion, which can moreover be directly tested by suitable instruments, that the tissues of the body act as condensers of considerable capacity.

CHAPTER VI.

COMBINATION OF CONDENSERS WITH SELF-INDUCTION.

THE cases already considered of a circuit bridged over at one or more pairs of points by condensers have not included the more complicated systems in which self-induction exists in any of the sections of the conductor. In practice a coefficient of self-induction would always exist in the section including the generator, and might be introduced into any of the other sections with the instruments inserted into them. It is therefore desirable to know what effect it would have upon the currents of the system. The general case may be represented by the following:—

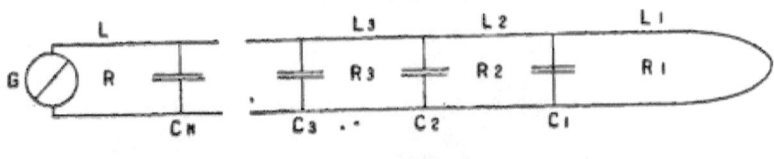

FIG. 14.

Here $C_1, C_2, C_3, \ldots C_n$, represent the capacities of the condensers, beginning from the remote end of the circuit.

$R_1, R_2, R_3, \ldots R_n, R$, the resistances of the sections.

$L_1, L_2, L_3, \ldots L_n, L$, the coefficients of self-induction of the sections, in the same order.

G represents the generator of the harmonically varying E.M.F.

Let T be half the period as usual.

To construct the proper diagram of electromotive forces, proceed as shown in Fig. 15.

Take any line EC to represent the effective E.M.F. in the section R_1.

Set off the angle CEQ in the direction of advancement, such that $\tan CEQ = \dfrac{L_1 \pi}{TR_1}$, and draw CQ perpendicular to EC.

Produce QE to M, so that, QE : EM :: $R_1 : R_2$.

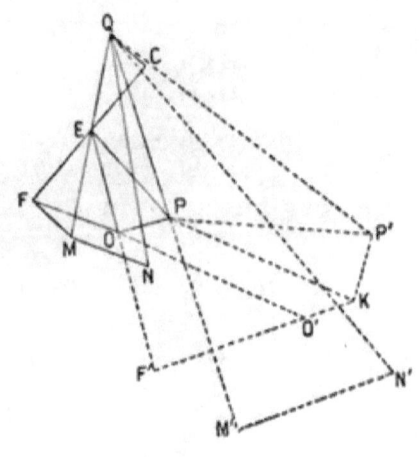

Fig. 15.

At Q set off the angle MQN in the direction of advancement, such that $\tan MQN = \dfrac{C_1 \pi R_1 R_2}{TR_1 + R_2}$, and draw MN at right angles to QM.

Draw MF at right angles to CE produced, and cutting it in F, thus making the two triangles, CEQ, FEM, similar.

From F draw FO parallel and equal to MN, and in the same direction as that line.

Join OE, and at E set off the angle OEP in the direction of advancement, such that $\tan OEP = \frac{L_2 \pi}{TR_2}$; draw OP perpendicular to EO, and join PQ.

Then PQ is the E.M.F. of the condenser C_2, or if there be but one condenser, it will be the E.M.F. of the generator.

EQ is the E.M.F. of the condenser C_1,

OE is the effective E.M.F. in the section R_2,

and EC, by hypothesis, that in the section R_1.

These points are proved as follows:—

Since EC is the effective E.M.F. in R_1, which possesses a coefficient of self-induction L_1, the impressed E.M.F. of this section must be such an E.M.F. as has its phase and magnitude determined as EQ has been obtained from EC, *i.e.*, it must be the hypothenuse of a right-angled triangle having the effective E.M.F. for a base, and the angle between the base and hypothenuse such that its tangent is equal to $\frac{L_1 \pi}{TR_1}$, and this angle must be in the direction of advancement from the base.

Hence EQ is the E.M.F. of the condenser C_1.

Since QE : EM :: R_1 : R_2, and the angle MQN has for its tangent, $\dfrac{C_1 \pi R_1 R_2}{TR_1 + R_2}$,

and NM = QM . tan MQN,

therefore NM = QM . $\dfrac{C_1 \pi R_1 R_2}{TR_1 + R_2}$.

Hence $\dfrac{NM}{R_2} = C_1 \pi \cdot \dfrac{R_1}{TR_1 + R_2} \cdot QM \cdot = \dfrac{C_1 \pi}{T} \cdot QE$,

and NM is at right angles to QE;

therefore $\dfrac{NM}{R_2}$ is the charging current of the condenser C_1, or NM is the E.M.F. which, acting through the resistance R_2, would produce the charging current for that condenser; and $NM = OF$. Hence $\dfrac{OF}{R_2}$ is the charging current. But $\dfrac{EC}{R_1}$ is the current in R_1 discharging the condenser in that section, and $\dfrac{EC}{R_1} = \dfrac{FE}{R_2}$.

Hence $\dfrac{EF}{R_2}$ is that current considered as one of the two components of the charging current $\dfrac{OF}{R_2}$.

Therefore $\dfrac{OE}{R_2}$ is the other component, *i.e.*, OE is the effective E.M.F. in the section R_2.

But this section has a coefficient of self-induction L_2, and $\tan OEP = \dfrac{L_2 \pi}{TR_2}$.

Therefore PE is the impressed E.M.F. necessary to produce the effective E.M.F. OE in this section.

But the impressed E.M.F. in R_2 is the resultant of the two E.M.F.s of the condensers C_2 and C_1, and EQ is the E.M.F. of the condenser C_1.

Therefore, finally, PQ is the E.M.F. of the condenser C_2, or, if there be but one condenser, C_1, PQ is the E.M.F. of the generator.

The extension of the diagram to give the E.M.F.s involved in the next section, R_3, will be easily understood if the steps are merely indicated, as follows:—

$QM' : QP :: R_3 + R_2 : R_2$,

$\tan M'QN' = \dfrac{C_2 \pi R_2 R_3}{TR_2 + R_3}$, $QM'N' = \dfrac{\pi}{2}$,

$OF' : EO :: R_3 : R_2$,

$F'O'$ is equal and parallel to $M'N'$,

$O'K$ is equal and parallel to OP,

$\tan KPP' = \dfrac{L_3 \pi}{TR_3}$, $\qquad PKP' = \dfrac{\pi}{2}$.

Then $P'Q$ is the E.M.F. of the condenser C_3, or, if there be but two condensers, of the generator, and KP is the effective E.M.F. in R_3, *i.e.*, $\dfrac{KP}{R_3}$ is the current in the section R_3.

Again, from the two lines $P'Q$, KP, with a further knowledge merely of the resistance and coefficient of self-induction of the next section R_4, we can deduce the current in that section and the E.M.F. of the generator or condenser C_4, as the case may be. Thus a complete diagram may be made showing the phases and magnitudes of every E.M.F. involved.

On contemplating Fig. 15 there is one important point to be noticed, viz., that self-induction in the sections by no means necessarily diminishes the currents in them, but up to a certain point may be actually beneficial. This cannot be the case when there is no capacity in circuit. Under such circumstances the self-induction must invariably diminish the current produced by a fixed electromotive force, but when condensers exist, it may happen that a smaller E.M.F. in the generator will produce a given heating effect in the sections of the circuit by virtue of the self-induction in the sections, than would have been required had there been no self-induction.

In the diagram, QO would have been the position of the line representing the E.M.F. of the condenser C_2 had there been no self-induction in the section R_2. As there is a coefficient of self-induction L_2 attaching to this section, the line of E.M.F. of the condenser C_2 takes the position QP. Now OP is merely drawn at right angles to OE, and such that $OP = OE \dfrac{L_2 \pi}{TR_2}$. It is clear, therefore, that when the angle QEO is less than 180°, a certain amount of self-induction will make QP smaller. This will be the case until QPO is a right angle, after which an increase in the coefficient of self-induction will necessitate a larger impressed E.M.F. to produce the given effect.

Similarly if QPK is less than 180°, QP′ may be diminished by increasing L_3 up to the point when QP′K is a right angle.

When QP′ is at a minimum, and therefore at right angles to KP′, it is parallel to PK. In other words, the phase of the impressed E.M.F. coincides with that of the current in the generating section. And similarly, for every section the E.M.F. of the condenser C_m would be a minimum when its phase corresponds to that of the current in R_m, a condition settled by the proper finite value being given to L_m.

An important case is that in which a generator operates in a circuit which is only closed by a condenser. If the condenser is a leaky one, the resistance through the leak is that represented by R_1, the remainder of the circuit is represented by R_2, for which there is a coefficient of self-induction L_2. Here L_1 is zero, and therefore in Fig. 15 the angle QEC will be zero, or Q will coincide with C, F with M, and O with N, and the diagram is reduced to Fig. 16.

CONDENSERS WITH SELF-INDUCTION.

When there is no self-induction, the impressed E.M.F. for given effective E.M.F.s in the two circuits is CO;

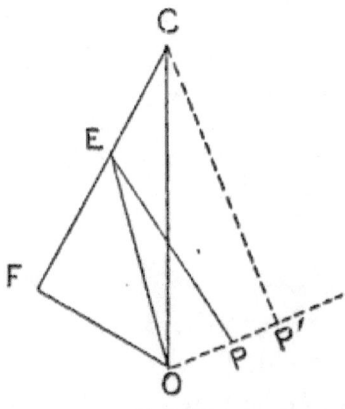

Fig. 16.

but as L_2 increases from zero the impressed E.M.F. necessarily becomes smaller, as CP, where $\tan PEO = \dfrac{L_2 \pi}{TR_2}$. This goes on until P arrives at P', when the angle CP'O is a right angle, at which point the impressed E.M.F. is a minimum. The condition that this shall be the case is easily deduced from the geometry of the figure, and is expressed by the equation

$$\frac{T^2}{\pi^2 C}\left\{\frac{1}{L_2} - \frac{1}{CR_1^2}\right\} = 1 \quad \ldots \ldots \quad (a)$$

which shows it to be independent of R_2, or the resistance in the section including the generator. From the above the proper value for L_2 may be deduced. It may, however, be observed, that as long as L_2 approaches the value indicated by this equation, all practical advantage will be gained, because, CP'O being a right angle, a variation in the value of the angle OEP' will only affect the length of CP' infinitesimally.

The proportion in which the presence of the beneficial coefficient of self-induction renders it possible to reduce the impressed E.M.F. is the relation which CP′ bears to CO. This relation is expressed by the fraction

$$-\frac{R_1 \sin^2 a + R_2}{\sqrt{(R_1 + R_2)^2 \sin^2 a + R_2^2 \cos^2 a}},$$

where a is the angle OCE, *i.e.*, is the angle whose tangent is equal to $\dfrac{C \pi R_1 R_2}{T R_1 + R_2}$.

The case is still simpler where the condenser is not a leaky one. In this case R_1 is infinite, and in Fig. 16 the point E coincides with F, and the line OE is coincident with the line OF, so that OP, which has to be at right angles to OE, becomes parallel to CF, the circumstances being simply represented by Fig. 17.

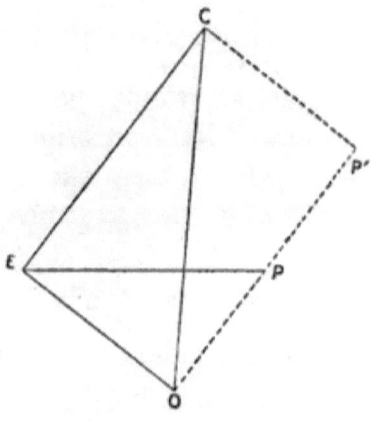

Fig. 17.

From this figure it is evident that when the self-induction coefficient is of the proper magnitude to allow the impressed E.M.F. to be a minimum, as represented by CP′, then this E.M.F. will also be the value of the effective E.M.F. in the circuit, represented by OE.

The tangent of OCE is $\dfrac{C\pi R_2}{T}$, which is the value of $\dfrac{C\pi R_1 R_2}{T \cdot R_1 + R_2}$, when R_1 is made infinite, and the tangent of PEO is $\dfrac{L\pi}{TR_2}$.

But when P moves to P′ these angles become complementary to each other; that is, the tangent of one of them is equal to the inverse of the tangent of the other.

Hence this condition is expressed by the equation

$$\dfrac{C\pi R_2}{T} = \dfrac{TR_2}{L\pi},$$

whence $T = \pi\sqrt{CL}$. . . . . (β)
R_2 disappearing from the equation.

This condition can also be deduced at once from the equation (α) by putting $R_1 = $ infinity.

The result may be summed up in the following words:—When an alternating generator operates upon a circuit which is closed by a condenser without leakage, and which possesses a coefficient of self-induction, then there is a certain period of alternation which may be given to the generator, at which the condenser might be replaced by a junction introducing no additional resistance into the circuit, the coefficient of self-induction being also removed, without disturbing the current. The condenser, in fact, in conjunction with the coefficient of self-induction, will obliterate the effects of the breach of continuity in the conductivity caused by the infinite resistance of the condenser itself. Moreover, this state of things is quite independent of the resistance of the circuit itself, which will then simply regulate the current in the same manner as with continuous uniform electromotive force.

It is to be observed that the advantage gained by establishing the critical relationship between the capacity, the coefficient of self-induction, the period, and the resistance in the remote section, when that is finite, is simply one of diminished potential. It is not one giving greater direct economy, because the power at work will be the same whether that relationship is established or not, since (Fig. 16) the power is equal to $\dfrac{OE}{2R_2} \times PC$ multiplied by the cosine of the angle between EO and PC, *i.e.*, the power is equal to $\dfrac{OE, P'C,}{2R_2}$ even if P does not coincide with P'. A similar remark may be made in other cases.

CHAPTER VII.

CONDENSER TRANSFORMER.

IT has been pointed out that the current in a circuit is modified, by attaching two points to the plates of a condenser, in such a way that there will actually be different currents in the two sections of the circuit. When there is no self-induction in the remote section of the circuit, the current in that section will always be less than that in the section containing the generator. But if there exists self-induction in the remote section, it may happen that the current in that section will be *greater* than the current in the section containing the generator. This effect will depend upon the capacity of the condenser, the coefficient of self-induction in the remote section, the resistance of the remote section, and the period. When these matters are related to each other in a proper way, we shall get a larger current through the remote section than through the generator, and the arrangement forms a veritable Transformer, which may be called a Condenser Transformer.

To examine the case, let us express one of the currents in terms of the other by the help of Fig. 15, dealing only with one condenser. Remembering that EC, OE, are the effective electro-motive forces in the two sections, the remote and the near respectively, and that R_1, R_2 are the resistances in those two sections, we must try to obtain the relation between $\frac{OE}{R_2}$ and $\frac{EC}{R_1}$, which currents may be called I_2 and I_1 respectively.

For brevity call the angle CEQ, β, and the angle MQN, γ.

So that $\tan \beta = \dfrac{L_1 \pi}{T R_1}$

and $\tan \gamma = \dfrac{C \pi R_1 R_2}{T(R_1 + R_2)}$

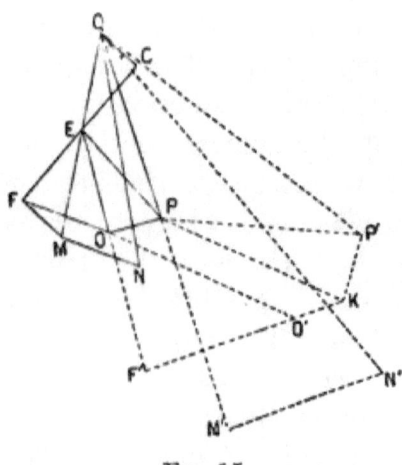

Fig. 15.

In the first place $OF = MN = QM \tan \gamma$
$$= \frac{R_1 + R_2}{R_1} . QE . \tan \gamma$$
$$= \frac{R_1 + R_2}{R_1} . \frac{EC}{\cos \beta} \tan \gamma$$

Secondly, $FE = \dfrac{EC . R_2}{R_1}$

Also $OE^2 = OF^2 + FE^2 - 2OF.FE \sin \beta$.

In this last equation substitute for OF and FE as found immediately above, thus :—

$$OE^2 = \left(\frac{R_1+R_2}{R_1}\right)^2 EC^2 \frac{\tan^2\gamma}{\cos^2\beta} + \left(\frac{R_2}{R_1}\right)^2 EC^2 - 2.\frac{\overline{R_1+R_2}R_2}{R_1^2}$$
$$EC^2 \tan\beta.\tan\gamma.$$

Divide through by R_2^2 and substitute I_2 for $\frac{OE}{R_2}$ and I_1 for $\frac{EC}{R_1}$

$$I_2^2 = I_1^2 \left\{ 1 + \left(\frac{R_1+R_2}{R_2}\right)^2 \frac{\tan^2\gamma}{\cos^2\beta} - 2\frac{R_1+R_2}{R_2}\tan\beta\tan\gamma \right\}$$

Clearing this expression of β and γ, we have for the relation between the squares of the currents,

$$\left(\frac{I_2}{I_1}\right)^2 = \left\{ 1 + \frac{C^2\pi^2 R_1^2}{T^2}\left(1 + \left(\frac{L_1\pi}{TR_1}\right)^2\right) - 2\frac{C\pi^2 L_1}{T^2} \right\}$$

This is the equation from which to draw conclusions as to the relative magnitude of the two currents. The following are among the most important.

(1) If there is no condenser the two currents are equal. Mathematically if $C = 0$ then $\frac{I_2}{I_1} = 1$.

(2) If there is no self-induction in the remote section a condenser will always cause the current in the generating section to be larger than in the remote section. If $L_1 = 0$, then $\left(\frac{I_2}{I_1}\right) > 1$.

(These two conclusions have been already established above.)

Writing the above equation somewhat differently,

$$\left(\frac{I_2}{I_1}\right)^2 = 1 + \frac{C\pi^2}{T^2}\left\{ CR_1^2\left(1 + \left(\frac{L_1\pi}{TR_1}\right)^2\right) - 2L_1 \right\}$$

we see further :—

(3) The relative value of the two terms
$$CR^2_1\left(1 + \left\{\frac{L_1\pi}{TR_1}\right\}^2\right) \text{ and } 2L_1$$

settles whether I_2 is greater than I_1. If the former is greater than the latter, then I_2 is greater than I_1, the current in the generating section greater than in the remote section, and *vice versâ*.

(4) The currents are equal when there is an equality between these two terms, *i.e.*, when

$$C = \frac{2L_1}{R_1^2\left(1 + \left(\frac{L_1\pi}{TR_1}\right)^2\right)}$$

(5) If we suppose that L_1, T, R, are invariable, and consider C as varying from zero upwards, I_2 is at first less than I_1, though when C is zero $I_2 = I_1$.

When C reaches the value $\dfrac{2L_1}{R_1^2\left(1 + \left(\frac{L_1\pi}{TR_1}\right)^2\right)}$

there is again equality between I_2 and I_1. Hence there must be some intermediate value of C which will make $\dfrac{I_2}{I_1}$ a minimum. The particular minimum value of $\dfrac{I_2}{I_1}$ occurs when $C = \dfrac{L_1}{R_1^2\left(1 + \left(\frac{L_1\pi}{TR_1}\right)^2\right)}$ which is half the value which would bring I_2 and I_1 into equality.

(6) In the case of $\dfrac{I_2}{I_1}$ being a minimum, which may be described as the case of maximum transformation upwards, *i.e.*, from small to large current, the ratio $\left(\dfrac{I_2}{I_1}\right)^2$ will have the value

$$\frac{1}{1 + \left(\frac{L_1\pi}{TR_1}\right)^2}$$

Thus the greater the coefficient of self-induction and

the less the resistance and the period, the greater will be our power to transform upwards by this means.

To give some idea of the magnitudes which might be involved, suppose a case where the period is the $\frac{1}{100}$th part of a second, the coefficient of self-induction ·04, and the resistance of the circuit beyond the condenser 14·5 ohms. Then a condenser of slightly under 48 microfarads capacity would cause the current in the remote section to be double that in the generator's section. Shortly after the publication of the first edition of this work, the Author successfully attempted upward transformation on the plan here indicated. The fact was communicated to a meeting of the Dynamic Society at the time.

CHAPTER VIII.

DISTRIBUTED CONDENSER.

The problem of the evenly distributed condenser differs somewhat from the foregoing cases of condensers situated at points of a circuit, in that the current in the conductor at different points is neither the same at two consecutive points, nor in the same phase. Each element of the conductor has its potential continually changing by the excess or defect of the flux of electricity from and to the point considered. Thus the phase of the potential, the phase of the current, the maximum values of both potential and current, will vary according to the point of the conductor considered. Mathematical analysis will serve, therefore, rather better than geometrical demonstration.

The differential equations which express the condition of flow at a point are easily obtained.

Let E be the potential at a point P,

$E + \delta E$ be the potential at a point Q, at distance δx from P,

ρ be the resistance of unit length of the conductor,

C be the capacity of unit length of the conductor,

I be the current at P,

$I + \delta I$ be the current at Q,

δt be an element of time, and δx of length.

Then $\rho\delta x$ is the resistance between P and Q; therefore $\dfrac{\delta E}{\rho \delta x} = -I$, the current being positive in the direction of the increase of x. Therefore, $\dfrac{dE}{dx} = -\rho I$ . . (a)

The difference between the current at Q and P has during the time δt diminished the potential of the element PQ. Since δE is the decrease of potential and $C \delta x$ is the capacity of the element, the quantity of electricity which has left the element PQ is $-C\delta x \cdot \delta E$, and must be equal to the balance of the electricity leaving and the electricity entering the element during the small interval δt.

$$\therefore \delta I \cdot \delta t = -C\delta x \cdot \delta E.$$

Therefore, ultimately

$$\dfrac{dI}{dx} = -C\dfrac{dE}{dt} \quad \ldots \quad (\beta)$$

By differentiating (a)

$$\dfrac{d^2 E}{dx^2} = -\rho \dfrac{dI}{dx}$$

and combining this with (β)

$$\dfrac{d^2 E}{dx^2} = \rho C \dfrac{dE}{dt} \quad \ldots \quad (\gamma)$$

Then these equations must be true in the complete solution of any particular case.

In the case of an infinitely long conductor, if subject to an impressed alternating E.M.F. ϵ whose period is 2 T, the time being measured from its zero phase when its value is increasing (or from the epoch of its ascending node), the solution is

$$E = \epsilon e^{-\sqrt{\frac{\rho C \pi}{2T}} x} \sin \frac{\pi}{T}\left(t - \sqrt{\frac{\rho CT}{2\pi}} x\right)$$

$$I = \epsilon \sqrt{\frac{C\pi}{T\rho}} \cdot e^{-\sqrt{\frac{\rho C \pi}{2T}} x} \sin \frac{\pi}{T}\left(t - \sqrt{\frac{\rho CT}{2\pi}} x + \frac{T}{4}\right) \bigg\} (\delta)$$

Here x is the distance from the point at which the alternating potential has its maximum value ϵ.

The proof of these formulæ rests upon their satisfying the equations (a), (β), (γ), and the condition that when $x = 0$, the maximum value of the potential is ϵ.

Among points worthy of notice in these formulæ, are (1) that at any point whatsoever the phase of the current precedes the phase of the potential by the same interval of time, viz., $\frac{1}{8}$ of the whole period of revolution, expressed by $\frac{T}{4}$.

(2) Since the phase of either potential value or current is continually postponed in proportion to x, we have a series of points equidistantly situated along the line which are in the same phase of their alternation.

This distance is equal to
$$2\sqrt{\frac{2\pi T}{C\rho}}$$
which may be called the wave length of the undulation.

The series of points bisecting this first series are in an opposite phase.

(3) Although the total resistance of an infinitely long conductor is infinite, still there may be a considerable flow of electricity at any point, if there be capacity distributed along the conductor.

The effect of the current on the dynamometer at any point, x, will be such that if I_1 is the direct current giving the same reading on the same instrument

DISTRIBUTED CONDENSER.

$$I_1 = \frac{1}{\sqrt{2}} \epsilon \sqrt{\frac{C\pi}{T\rho}} e^{-\sqrt{\frac{\rho C\pi}{2T}} \cdot x}.$$

The reading of the dynamometer will be—

$$\epsilon^2 \frac{C\pi}{2T\rho} \cdot e^{-\sqrt{\frac{\rho C\pi}{2T}} 2x}.$$

or, when $x=0$, $\epsilon^2 \dfrac{C\pi}{2T\rho}$, which is equal to $\dfrac{\epsilon^2}{2\left(\dfrac{T\rho}{C\pi}\right)}$.

By the reading of the dynamometer is meant the angle of the index on a uniform graduation in which the unit angle is that indicated when a direct uniform unit current is passing through the instrument, the coils being in series. Under these circumstances, when an harmonic current is passing through the instrument, of maximum value I, then the reading will be $\dfrac{I^2}{2}$.

A uniform current, I, would of course produce a reading I^2.

In what follows, therefore, the dynamometer reading may be taken to mean the value of half the square of the maximum value of an harmonic current.

The equations (δ) give the values of the potential and the current at any point of the conductor and at any time. They indicate that these quantities vary harmonically, and the factors which are outside the periodic function of the time give the maximum values of the quantities.

If we were to divide the maximum value of the potential by the maximum value of the current, we should of course get something of the order of resistance, which might in a certain sense be called the equivalent resistance. Performing this operation, the resistance will be found to be in every case $\sqrt{\dfrac{T\rho}{C\pi}}$.

Call this quantity B; it will be wanted in comparison later on.

Since, however, the phase of the E.M.F. lags behind that of the current by ⅛th of the period, it would be instructive to make the case parallel to one in which a generator of E.M.F. equal to that at any point considered plays into a circuit of resistance r, closed by a perfectly non-leaking condenser of capacity K, and endeavour to find what values must be given to r and K, so that the current may be the one at the point considered in phase and magnitude. Then r and K may be more properly called the equivalent resistance and the equivalent capacity of the infinitely long circuit.

The necessary values are found thus:—

Since the current precedes the E.M.F. by one-eighth of the period, the geometrical diagram of the E.M.F.s concerned will be a right-angled triangle, having each of the smaller angles equal to $\dfrac{\pi}{4}$.

Hence $\dfrac{K\pi r}{T} = \tan\dfrac{\pi}{4} = 1$, $\therefore K = \dfrac{T}{\pi r}$.

Again, the effective E.M.F. bears to the impressed E.M.F. the relation of $1 : \sqrt{2}$, as gathered from the diagram; and the effective E.M.F. divided by the resistance is equal to the current.

Calling a the impressed E.M.F., *i.e.*, the maximum value of E,

β the effective E.M.F.,

γ the current, *i.e.*, the maximum value of I,

for the sake of putting the above statements in algebraical form, we have,

$$\frac{\beta}{a} = \frac{1}{\sqrt{2}}, \qquad \frac{\beta}{r} = \gamma.$$

Hence $r = \dfrac{\beta}{\gamma} = \dfrac{a}{\sqrt{2}.\gamma} = \dfrac{1}{\sqrt{2}} \dfrac{\text{maximum value of E}}{\text{maximum value of I}}$.

$$\left. \begin{aligned} \therefore \quad r &= \frac{1}{\sqrt{2}} \sqrt{\frac{T\rho}{C\pi}} \\ \text{and} \quad K &= \frac{T}{\pi} \frac{\sqrt{2}.\sqrt{C\pi}}{\sqrt{T\rho}} = \sqrt{\frac{2CT}{\pi\rho}} \end{aligned} \right\} \quad \cdots \cdots (\epsilon)$$

Thus the equivalent resistance varies as the square root of the specific resistance, and inversely as the square root of the specific capacity. The equivalent capacity varies as the square root of the specific capacity and inversely as the square root of the specific resistance. Both quantities vary as the square root of the period. Their product is equal to $\dfrac{T}{\pi}$, depending merely on the period.

We have now to consider the cases where the conductor is not infinitely long, but has a length l.

The two important cases are—

(1) When the distant end is put to earth, and is kept, therefore, at zero *potential*.

(2) When the distant end is perfectly insulated, and, therefore, is subject to zero *current* at any instant of time.

These cases are in reality very similar to each other.

Suppose, as before, that the conductor is infinitely long, and that at a point $2l$ distant from the zero point the source of another alternating E.M.F. is inserted, whose maximum value and period are ϵ and $2T$ respectively, as in the primary case, but whose phase is *exactly opposite* to the E.M.F. at the zero point. It is clear, then, that the point l from the origin, since it is equidistant from both sources, has its potential raised as much by one alternation as depressed by the other, and the potential will remain, therefore, at zero. The point may be put to earth. For the potential and current at any point between 0 and l we have merely to add the effects of the two alternations. We have already expressions for the primary source. Calling those due to the secondary source *reflected* potentials and currents, we have as follows:—

The reflected potential at any point and time will be

$$\epsilon\, e^{-\sqrt{\frac{\rho C \pi}{2T}}(2l-x)} \sin\frac{\pi}{T}\left\{ t - \sqrt{\frac{\rho C T}{2\pi}} \cdot (2l-x) + T \right\},$$

and the reflected current, positive towards the origin of x

$$\epsilon\sqrt{\frac{C\pi}{T\rho}} \cdot e^{-\sqrt{\frac{\rho C \pi}{2T}} \cdot (2l-x)} \sin\frac{\pi}{T}\left\{ t - \sqrt{\frac{\rho C T}{2\pi}}(2l-x) + \frac{T}{4} + T \right\}$$

Combining the original effect and the reflected effect, and writing for brevity $\sqrt{\frac{\rho C \pi}{2T}} = a$, we have for the potential at any time and place—

$$E = \epsilon e^{-ax}\left\{ 1 + e^{-4a(l-x)} - 2e^{-2a(l-x)}\cos 2a(l-x) \right\}^{\frac{1}{2}} \sin\left\{ \frac{\pi t}{T} - ax + y \right\},$$

where
$$\tan y = \frac{e^{-2a(l-x)}\sin 2a(l-x)}{1 - e^{-2a(l-x)}\cos 2a(l-x)},$$

which shows that the phase of the potential is advanced by the finite conductor being put to earth, compared with the phase at the corresponding point of an infinite conductor.

The current at any point of time or place I is given by

$$I = \epsilon \sqrt{\frac{C\pi}{T\rho}}\, e^{-ax} \left\{ 1 + e^{-4a(l-x)} + 2 \cdot e^{-2a(l-x)} \cos 2a(l-x) \right\}^{\frac{1}{2}} \sin\left\{ \frac{\pi t}{T} - ax + \frac{\pi}{4} - z \right\},$$

where

$$\tan z = \frac{e^{-2a(l-x)} \sin 2a(l-x)}{1 + e^{-2a(l-x)} \cos 2a(l-x)},$$

which shows that the phase of the current is delayed by the finite conductor being put to earth, compared with the phase at the corresponding point of an infinite conductor. We cannot, however, say that the current and E.M.F. at any point differ in phase always by the same amount. The amount varies at different points.

These formulæ enable a comparison to be made of the squares of the currents, as indicated by a dynamometer, at different points of a line. The dynamometer readings will be as before $\frac{(\text{maximum I})^2}{2}$, that is to say,

$$\epsilon^2 \frac{C\pi}{2T\rho} \cdot e^{-2ax} \left\{ 1 + e^{-4a(l-x)} + 2 \cdot e^{-2a(l-x)} \cos 2a\overline{l-x} \right\}$$

This, when $x = 0$, is equal to

$$\epsilon^2 \frac{C\pi}{2T\rho} \cdot \left(1 + e^{-4al} + 2 \cdot e^{-2al} \cos 2al \right)$$

and, when $x = l$, it has for value

$$\epsilon^2 \frac{C\pi}{2T\rho} \cdot e^{-2al} 4.$$

The relation of the former to the latter is therefore

$$\frac{1 + e^{-4al} + 2e^{-2al}\cos 2al}{4e^{-2al}}$$

$$= \tfrac{1}{4}\left\{e^{2al} + e^{-2al} + 2\cos 2al\right\} \quad . \quad . \quad . \quad (\epsilon)$$

Which is equal to the series

$$1 + \frac{(2al)^4}{\underline{/4}} + \frac{(2al)^8}{\underline{/8}} + \; . \; .$$

and, therefore, always more than unity, however small l may be. Hence the dynamometer reading is a minimum at $x = l$.

This case corresponds to that of an alternating current machine in connection with a uniform cable possessing capacity. The point $x = l$ corresponds to a point of the cable half way between the terminals of the machine, at which point the potential is zero always, and ϵ corresponds to half the E.M.F. of the machine.

The comparative reading of the dynamometer at any point distant l from the central point in either direction will be

$$1 + \frac{(2al)^4}{\underline{/4}} + \frac{(2al)^8}{\underline{/8}} + \text{etc.}$$

It is clear, therefore, that three readings of the dynamometer at points at measured distances apart from one another will suffice to determine the value of a, and $a = \sqrt{\rho\frac{C\pi}{2T}}$; therefore, if ρ, the rate of resistance, and $2T$, the periodic time, are known, C, the capacity of a unit of length of the conductor, can be accurately determined.

If we employ the symbols of hyperbolic trigonometry the expression (ϵ) may be written

$$\frac{\cosh 2al + \cos 2al}{2};$$

that is to say, the relation between the dynamometer readings at any point distant l from the central point or point of no potential, and at that central point is expressed by the mean of the hyperbolic and circular cosines, a result given before the Society of Telegraph-Engineers and Electricians by the Author.

If there had been an ordinary conductor possessing no electrical capacity, but of such a resistance, B, that an alternating E.M.F. corresponding to the potential existing in the condenser cable at a point, viz.:—

$$\epsilon \cdot e^{-ax}\left\{1 + e^{-4a(l-x)} - 2e^{-2a(l-x)}\cos 2a(l-x)\right\}^{\frac{1}{2}}$$

would produce the same effect on the dynamometer as is observed, then

$$\frac{\epsilon^2 e^{-2ax}\left\{1 + e^{-4a(l-x)} - 2e^{-2a(l-x)}\cos 2a\overline{l-x}\right\}}{2B^2}$$

$$= \frac{\epsilon^2 C\pi}{2T\rho}e^{-2ax}\left\{1 + e^{-4a(l-x)} + 2e^{-2a(l-x)}\cos 2a\overline{l-x}\right\}$$

We deduce

$$B^2 = \frac{T\rho}{C\pi}\left\{\frac{1 + e^{-4a(l-x)} - 2e^{-2a(l-x)}\cos 2a(l-x)}{1 + e^{-4a(l-x)} + 2e^{-2a(l-x)}\cos 2a\ \overline{l-x}}\right\}$$

When $x = 0$,

$$B = \sqrt{\frac{T\rho}{C\pi}} \cdot \left\{\frac{1 + e^{-4al} - 2e^{-2al}\cos 2al}{1 + e^{-4al} + 2e^{-2al}\cos 2al}\right\}^{\frac{1}{2}}$$

If $2al = \theta$, for brevity

$$B = \sqrt{\frac{T\rho}{C\pi}} \cdot \left\{ \frac{\frac{\theta^2}{\underline{/2}} + \frac{\theta^6}{\underline{/6}} + \frac{\theta^{10}}{\underline{/10}} + \cdots}{1 + \frac{\theta^4}{\underline{/4}} + \frac{\theta^8}{\underline{/8}} + \cdots} \right\}^{\frac{1}{2}}$$

When $\theta = 0$ the term in the bracket is equal to zero, when $\theta = \infty$ it is equal to unity.

Hence B increases from zero to $\sqrt{\frac{T\rho}{C\pi}}$, as l increases from zero to infinity.

Thus it will be seen how little the virtual resistance, as far as the dynamometer at the generating station is concerned, is increased by increasing the length of the cable. In fact, the virtual resistance tends to a limiting value $\sqrt{\frac{T\rho}{C\pi}}$.

The case of the open circuit differs from the last in that the source which produces the reflected current and potential is in the same phase as the primary source, so that the currents at the middle point neutralize each other. As in such a case no current passes across the middle point, the variation of the potentials along the line is entirely due to the propagation of elasticity at the primary source; the conductor may be cut at the middle point and the end treated as free and perfectly insulated, a point of no current, as in the former case it was a point of no variation in potential.

The *reflected* potential will be

$$\epsilon\, e^{-\sqrt{\frac{\rho C\pi}{2T}}(2l-x)} \sin \frac{\pi}{T}\left\{t - \sqrt{\frac{\rho C T}{2\pi}}(2l-x)\right\},$$

and the reflected current, positive towards origin of x,

$$\epsilon \sqrt{\frac{C\pi}{T\rho}} e^{-\sqrt{\frac{\rho C\pi}{2T}}(2l-x)} \sin\frac{\pi}{T}\left\{t - \sqrt{\frac{\rho CT}{2\pi}}(2l-x) + \frac{T}{4}\right\}$$

Combining the primary and secondary effects, and writing as before $\sqrt{\frac{\rho C\pi}{2T}} = a$, we have for potential at any time and place

$$E = \epsilon e^{-ax}\left\{1 + e^{-4a(l-x)} + 2e^{-2a(l-x)}\cos 2a(l-x)\right\}^{\frac{1}{2}}$$
$$\sin\left(\frac{\pi t}{T} - ax - y\right),$$

where
$$\tan y = \frac{e^{-2a(l-x)}\sin 2a(l-x)}{1 + e^{-2a(l-x)}\cos 2a(l-x)},$$

which shows that the phase of the potential is *postponed* by the *finite insulated* conductor, compared with the phase at the corresponding point of an *infinite* conductor.

The current at any place and point of time is given by

$$I = \epsilon\sqrt{\frac{C\pi}{T\rho}} e^{-ax}\left\{1 + e^{-4a(l-x)} - 2e^{-2a(l-x)}\cos 2a(l-x)\right\}^{\frac{1}{2}} \sin\left(\frac{\pi t}{T} - ax + \frac{\pi}{4} + z\right),$$

where
$$\tan z = \frac{e^{-2a(l-x)}\sin 2a(l-x)}{1 - e^{-2a(l-x)}\cos 2a(l-x)},$$

which shows that the phase of the current is *advanced* by the *finite insulated* conductor, compared with the phase at the corresponding point of an infinite conductor. But in this case also the difference of the phases of E.M.F. and current differs at different points of the circuit.

The dynamometer reading for any point will be

$$\epsilon^2 \frac{C\pi}{2T\rho} e^{-2ax} \left\{ 1 + e^{-4a(l-x)} - 2e^{-2a(l-x)} \cos 2a(l-x) \right\}$$

If we write $l-x=d$, and therefore $x=l-d$, we have for the dynamometer readings at distance d from the free end

$$\epsilon^2 \frac{C\pi}{2T\rho} e^{-2a(l-d)} \left\{ 1 + e^{-4ad} - 2e^{-2ad} \cos 2ad \right\}$$

$$= \epsilon^2 \frac{C\pi}{2T\rho} e^{-2al} e^{2ad} \left\{ 1 + e^{-4ad} - 2e^{-2ad} \cos 2ad \right\}$$

$$= \epsilon^2 \frac{C\pi}{2T\rho} e^{-2al} \left\{ e^{2ad} + e^{-2ad} - 2\cos 2ad \right\}$$

Which may be written

$$\epsilon^2 \frac{C\pi}{T\rho} e^{-2al} \left\{ \cosh 2ad - \cos 2ad \right\}$$

As in the former case, the virtual resistance B may be found by dividing the maximum value of E by the maximum value of I; whence,

$$B = \sqrt{\frac{T\rho}{C\pi}} \left\{ \frac{1 + e^{-4a(l-x)} + 2e^{-2a(l-x)} \cos 2a(l-x)}{1 + e^{-4a(l-x)} - 2e^{-2a(l-x)} \cos 2a(l-x)} \right\}^{\frac{1}{2}}$$

In this expression put $x=0$ and (for brevity) $2al=\theta$. Then B becomes

$$\sqrt{\frac{T\rho}{C\pi}} \left\{ \frac{1 + \dfrac{\theta^4}{\underline{4}} + \dfrac{\theta^8}{\underline{8}} + \ldots}{\dfrac{\theta^2}{\underline{2}} + \dfrac{\theta^6}{\underline{6}} + \dfrac{\theta^{10}}{\underline{10}} + \ldots} \right\}^{\frac{1}{2}}$$

If $\theta = 0$, *i.e.*, when the length of the conductor l vanishes, the expression in the bracket is infinite. But if θ is infinite, *i.e.*, when the length of the conductor is infinite, the expression in the bracket becomes equal to unity. Thus B diminishes from infinity to $\sqrt{\dfrac{T\rho}{C\pi}}$ as l increases from zero to infinity. Thus, in every case, $\sqrt{\dfrac{T\rho}{C\pi}}$ is the virtual resistance of an infinitely long cable, whether put to the earth or not.

The expressions for B may be made extremely simple by using the symbols of hyperbolic trigonometry. In the case of the conductor put to earth at distance l from the point of observation, we have

$$B_1 = \sqrt{\dfrac{T\rho}{C\pi}} \left\{ \dfrac{\cosh 2al - \cos 2al}{\cosh 2al + \cos 2al} \right\}^{\frac{1}{2}}$$

and in the case of the conductor having a free end at distance l from the point of observation.

$$B_2 = \sqrt{\dfrac{T\rho}{C\pi}} \left\{ \dfrac{\cosh 2al + \cos 2al}{\cosh 2al - \cos 2al} \right\}^{\frac{1}{2}}$$

CHAPTER IX.

DISTRIBUTED CONDENSER (*continued*)—TELEPHONY.

By the help of the formulæ which have been given, especially of those connected with the case in which the cable is put to earth, we can deal with some important questions in long distance telephony. The falling off in current corresponds to the failing in intensity of tone produced. The fact that this falling away in current depends upon the period, being greater as the period is smaller, will explain the alteration in character of a composite tone, those components which have a higher pitch suffering more decay than lower notes.

Table A, in the Appendix, will facilitate the calculation of such effects. As an example of the use of the table, suppose the following problem :—

A cable possessing ·2 microfarad of capacity per kilometre, and a resistance of $2\frac{1}{2}$ ohms per kilometre, is 40 kilometres long. What will be the relation of the current at the sending end to that at the receiving end for a tone whose period is $\frac{1}{500}$ of a second?

Adopting quadrant units, we find that since the cable has ·2 microfarad per kilometre, it has ·002 farad in one quadrant; therefore, $C = ·002$.

Since the cable possesses 2·5 ohms in one kilometre, it has 25,000 ohms in one quadrant; hence $\rho = 25,000$. The period is $\frac{1}{500}$ of a second; hence $2T = \frac{1}{500}$.

Since the length of the cable is 40 kilometres, $l = \cdot 004$ of a quadrant; hence a, which is equal to $\sqrt{\rho \dfrac{C\pi}{2T}}$, has for value $\sqrt{25,000 \times \cdot 002 \times \pi \times 500}$; or $a = 280 \cdot 25$, and $2\,a\,l = 2 \cdot 24$, which is the number corresponding to θ in the table.

Looking, therefore, horizontally between the lines headed 2·2 and 2·3 and in the column headed $\dfrac{\cosh \theta + \cos \theta}{2}$, we shall be able to deduce the number 2·06 by interpolation. This number is the relation of the square of the current at the sending place to the square of the current at the receiving place. The relation of the currents themselves will be the square root of this number, *i.e.*, about $\tfrac{144}{100}$.

Now, suppose we investigate what takes place with the note one octave lower than that already considered. In this case, $2\,T = \dfrac{1}{250}$, and a will be found to be 198·17, and $2\,a\,l$ will be 1·59, the number corresponding to θ in the table.

The value of $\dfrac{\cosh \theta + \cos \theta}{2}$, for this value of θ, will be found from the table to be 1·265, and the square root of this is 1·125, so that for this note the telephone current is reduced in the ratio of 112 : 100. For higher notes, the falling off of the currents would increase very rapidly. At the octave above the first note considered, the current would be reduced more than in the ratio 300 : 100.

Now the notes considered come well within the range of human tones in speech. The fundamental tones of the voice are, moreover, richly accompanied by harmonics of a high order, by means of which, however unconsciously, we interpret under ordinary circumstances the sounds received upon the ear. All such harmonics would suffer very materially in the course of the transmission.

Thus, at the end of a cable of any considerable length and capacity the various tones of the voice would be received in a state of degradation depending upon their pitch. If this were not the case, if all the tones were reduced in strength in the same proportion, a relay might be employed to restore the various currents to their original intensity, or to one in which the ear would readily appreciate the meaning of the tone. But the ear has not the synthetic power of reconstructing a composite tone from the wreck of variously degraded components. In this consideration reside the limits of telephony. And until it is more clearly understood than it seems to be at present, people will fail to understand the exquisite nonsense to which they are often now content to listen about the possibilities of being able to catch the minutest modulations of voice of a trans-oceanic prima donna, and so on. It would be more to the purpose to endeavour to keep $2 a l$ small. Lord Rayleigh at Montreal looked to alternating currents "to educate so-called practical electricians whose ideas do not easily rise above ohms and volts." It is to be hoped that this anticipation will be realized, for unfortunately the imputation conveyed is too well deserved. In a book specially published by the Institution of Civil Engineers in 1884, upon the practical Applications of Electricity, there is a paper by Sir Frederick Bramwell on Telephones, in which the following sentence occurs:—

"You all know that if an iron plate . . . be caused to approach towards or recede from a permanent magnet, that magnet being surrounded by a coil of insulated wire, there will be set up in the coil of wire electric currents, which will vary in direction according as to whether the plate is approaching the magnet or is receding from it."

In this sentence it is clearly assumed that the induction, whose phase accompanies the phase of the plate's motion, is also accompanied in phase by the current. The mistake is made, that induction is of the nature of current, instead of electromotive force. The induction produced by the plate's motion is the impressed electromotive force, but there may arise electromotive forces of self-induction, and those due to capacity in the conductor, which would cause the current to have a phase differing from that of the plate's motion, and to suffer at the same time a diminution in strength, both effects depending upon the pitch of the note producing the vibrations of the plate.

These are the effects which will ultimately give a limit to telephony. But farther on in the same paper we read, "As many persons have asked, What is the limit of possibility? I would say, that I should think it depends practically on the excellence of the insulation and the avoidance of induced currents."

Now, if by induced currents is meant currents produced by accidental induction from outside, it seems likely that such currents would not so materially affect the interpretation of the sound as would the degradation of the currents which it is desired should be transmitted. For the ear, which is ultimately the organ to interpret, is every day practising the art of discriminating between two or more contemporaneous and superimposed sounds. The ear is, in fact, an excellent analyser, but for the process of recomposition necessary before a degraded composite tone can be truly interpreted something more than power of analysis is necessary, and this the ordinary ear does not possess.

To return to investigation. The dynamometer reading at distance x from the generator in a cable put to earth at the distant end is

$$\epsilon^2 \frac{C\pi}{2T\rho} e^{-2ax} e^{-2a(l-x)} \left\{ e^{2a(l-x)} + e^{-2a(l-x)} + 2\cos 2a(l-x) \right\}$$

$$= \epsilon^2 \frac{C\pi}{T\rho} e^{-2al} \left\{ \cosh 2a(l-x) + \cos 2a(l-x) \right\}$$

$$= \epsilon^2 \frac{C\pi}{T\rho} e^{-2al} \left\{ \cosh 2ad + \cos 2ad \right\};$$

where d is the distance from the remote end, and for a cable whose end is insulated, it is

$$\epsilon^2 \frac{C\pi}{T\rho} e^{-2al} \left\{ \cosh 2ad - \cos 2ad \right\}.$$

The relation is $\dfrac{\cosh 2ad + \cos 2ad}{\cosh 2ad - \cos 2ad}$.

Now, a reference to the second and third columns of the table shows that the numerator of this fraction does not always exceed the denominator. At certain values of $2ad$, those in fact for which $\cos 2ad$ is equal to zero, and which recur periodically, the numerator is equal to the denominator, and the excess changes sides. The first of such cases occurs between $\theta = 1.5$ and $\theta = 1.6$, i.e., when $\theta = \dfrac{\pi}{2} = 1.57$.

If, therefore, the length and structure of the cable and the period of alternation are such that $2al$ is greater than 1·57, there will actually be regions of the cable at which, as far as the intensity of the currents received

DISTRIBUTED CONDENSER—TELEPHONY.

is concerned, it would be better if the circuit were an open one than that the distant end should be put to earth; and as we increase the length of the cable this state of things alternates with the other at fixed distances from the far end in the following way:—

Deduce a distance λ such that $2a\lambda = \dfrac{\pi}{2}$;

whence $$\lambda = \frac{\pi}{4a} = \frac{1}{4}\sqrt{\frac{2\mathrm{T}\pi}{\rho\mathrm{C}}},$$

and set off from the far end distances corresponding to λ, 3λ, 5λ, etc.

Then in the region within λ of the far end the current will be greater when the end is put to earth. In the region, length 2λ, between 3λ and λ from the far end, the current will be greater when the end is insulated. In the next region of length 2λ, it is best to put the far end to earth. And so on alternately for every length of 2λ. At the points where these alternating regions join, there will be the same current under both circumstances.

This distance 2λ is one quarter of the wave length of the undulation in the infinite conductor (*vide* p. 60).

For the cable and note given in the example, the length λ is 28 kilometres.

CHAPTER X.

THE TRANSMISSION OF POWER.

THE problem of the transmission of power by means of harmonic currents admits of being readily treated by the geometrical method, in the same fashion as local condensers and induction problems, and, moreover, some interesting particulars are rendered so plain by the use of geometry as almost to be self-evident.

To give a preliminary notion of the possibility of such an action, imagine an ordinary galvanometer subjected to alternating currents. The impulses on the needle considered as fixed would be as often and as strongly in one direction as in the other. But suppose that after an impulse in one direction the second and contrary impulse is not given *until the needle has passed over the dead point position,* which will be at right angles to the coils. In this case the second impulse will only add to the speed of the needle's rotation. Postpone the third impulse similarly until the second dead point has been passed, and again there will be positive acceleration. This elementary form of the problem enables the possibility of the motion to be conceived. But it does more than this, it shows that the direction of motion of the *receiving* machine is indeterminate. The impulse must always tend to give the needle one direction of rotation; but the direction simply depends upon which side of the dead point the needle may happen to be. Electric transmission of powers admits, therefore, of the following classification:—

A magneto machine giving a direct current and employed as a driver gives a current varying with its direction.

A similar magneto machine used as a follower takes a direction varying with that of the current.

A dynamo electric machine giving a direct current and used as a driver gives a current whose direction is indeterminate.

A dynamo electric machine used as a follower takes a direction determinate and independent of that of the current.

An alternating machine, whether used as a driver or a follower, may run in either direction indifferently. The period of the follower coincides with that of the driver.

To treat the question more generally and exactly, it must be remembered that in a simple circuit, possessing no capacity, and subject to sources of alternating electromotive force of the same period, we may combine all the electromotive forces into one having that period. Thus if AB, BC, CD, DE, EF, are a number of lines representing in magnitude and phase the various electromotive forces, and AF be joined, then AF is the single electromotive force which will produce the effects observed. If there exist self-induction or mutual induction the diagram can be completed in the manner already indicated in earlier articles, using AF as the single impressed E.M.F. We shall thus obtain a line representing the effective E.M.F., which will be in the same phase as the current. Dividing this by the resistance we can compute the current; and if we multiply the result by the projection upon this line of the line representing any of the individual component electromotive forces, and divide by 2, we shall obtain a measure of the power at work in the particular corresponding source. Now the projection of any individual line will be positive if the angle between it and the line of effective E.M.F. be less than a right angle, and negative if this angle be greater than a right angle.

In the former case the power is positive, and the source does work.

In the second case the power is negative, and the source has work done upon it.

Thus, in Fig. 18, let AB, BC, CD be the electromotive forces of three sources, laid down in their respective phases. Then AD is the resultant E.M.F.

The angle EDA has its tangent equal to $\frac{L\pi}{TR}$,

where L is the coefficient of self-induction of the circuit,

R is the resistance,

and T is the half period of the undulation.

AE is at right angles to DE.

From C and B, CC', BB' are drawn perpendicular to ED.

The power exerted by source of AB is therefore $\frac{ED}{2R} \cdot EB'$;

similarly the power exerted by source of CD is $\frac{ED}{2R} \cdot C'D$.

But the power of the source of BC is seen in the diagram to be negative. Hence this source has work done upon it with a power equal to $\frac{ED}{2R} \cdot C'B'$.

Such a source would constitute what is called a

motor; *i.e.*, it may be loaded in the mechanical sense of the word, and made to do work.

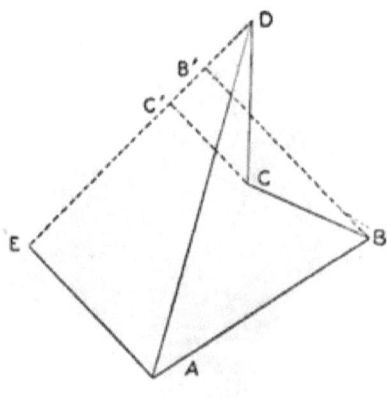

Fig. 18.

It will be seen that had the coefficient of self-induction been small, and E nearly coincided with A, the projection of BC upon ED would have been positive, and the source of BC would not have been a *recipient* source, but would have had to do part of the work of heating the circuit.

Having settled these preliminary ideas, let us suppose a case where the sources are two in number, represented in magnitude and phase by AB, BC (Fig. 19), of which BC is the smaller.

Upon AB as diameter describe a circle, and join AC and produce it to cut the circle in D.

From CA in the direction of retardation set off the angle CAF, having a tangent equal to $\frac{L\pi}{TR}$, the line AF cutting the circle in F.

L, T, and R, have the usual significations.

Draw CE perpendicular to AF.

Join BD, BF, which lines will be at right angles to AD, AF, respectively.

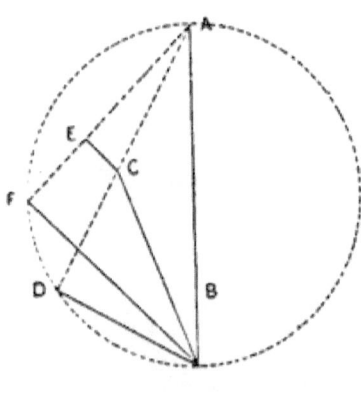

Fig. 19.

Then, calling for brevity the source of AB the source A, and the source of BC the source C, we have—

$$\text{The power exerted by Source A} = \frac{AE \cdot AF}{2R}$$

$$\text{The power-doing work on Source C} = \frac{AE \cdot FE}{2R}$$

Thus work will be done upon C so long as E lies within the semicircle AFB, upon which F always lies. But if the coefficient of self-induction is sufficiently increased, the angle DAF will open out, and F will move along the circle towards A until FCB are in one straight line. This condition obtains when L is large enough to satisfy the equation,

$$\frac{L\pi}{TR} \sin \theta = \cos \theta - \frac{f}{e},$$

where θ is the angle ABC, and $AB = e$, $BC = f$.

If BC is so great that the point C approaches the circumference, C will for that reason alone cease to be a recipient source, for the projection of BC upon the line of effective E.M.F. will either vanish, if the coefficient of self-induction is evanescent, or be positive if this latter has a finite value, in which case the source C does work.

But suppose that matters are represented actually by Fig. 19, in which the source C is, suppose, loaded to the extent represented by $\frac{FE \cdot EA}{2R}$. There will be a condition of equilibrium, but it is necessary to show that such a position will represent *stable* equilibrium; that is to say, it is necessary for steady motion that, if the angle of phase difference CBA be slightly varied, there shall be a restitution of the exact position and not an augmentation of the displacement.

For the consideration of this point we may suppose the phase difference to be the only variable element, the two electromotive forces and the coefficient of self-induction remaining constant.

With a variation of the angle ABC, C evidently describes a circle round B as centre. Now AE always bears the same relation to AC, viz., the cosine of the angle EAC, which remains always the same, since the coefficient of self-induction does so. Therefore, whatever curve C describes round A, E will describe an exactly similar one on a scale diminished as the cosine of the angle EAC.

The curve which E describes will therefore be a circle whose radius is equal to CB cos EAC.

The two circles described by C and E will not, however, be similarly situated with regard to AB, but since EA makes a constant angle with CA, the circle described by E will be displaced in the direction of retardation by the amount of the angle EAC, at the same time that it is diminished, and drawn nearer to A in the proportion of the cosine of this angle.

We thus arrive at the following construction, which to avoid confusion of lines we employ in a new figure, viz., Fig. 20. AB is the electromotive force e of the driving machine. BM is cut off from AB, so that BM

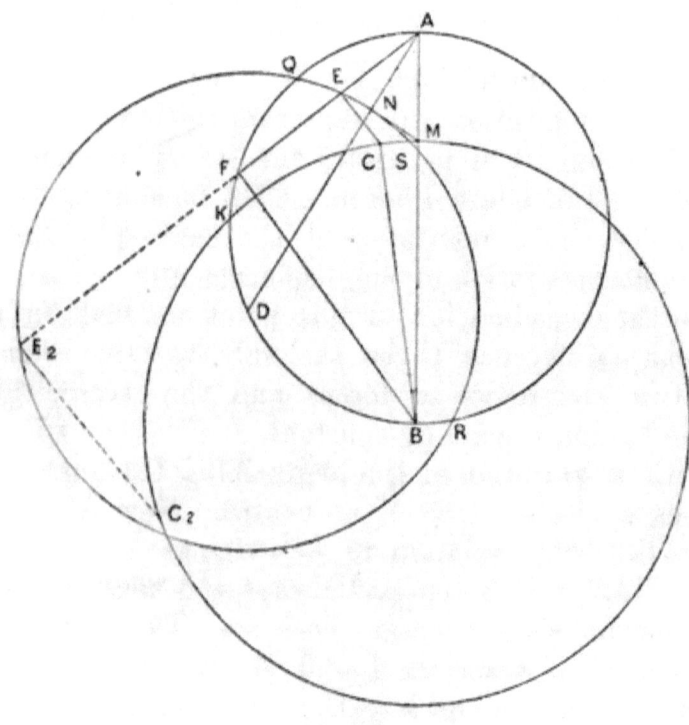

Fig. 20.

represents the electromotive force f of the follower in *magnitude* only. MCK is a circle described with BM as radius. BDK is a circle described on AB as diameter. BAD is the angle whose tangent is equal to $\frac{L\pi}{TR}$, and is set off from BA in the direction of retardation. MN is a perpendicular upon AD, so that $DN = BM \cos DAB$; SNEQ is a circle described with centre D and radius DN, cutting the circle BDK in Q, and R and MB in S. Then, since $AN : AM :: ND : MB :: $ cosine $DAB : 1$, and $\tan BAD = \dfrac{L\pi}{TR}$, from the foregoing considerations the circle SNEQ is the circle described by the point E of Fig. 19.

Now, let BC be some particular position of the E.M.F. of the follower, and E the corresponding point upon the circle whose centre is D.

AE produced cuts the circle BDA in F.

Then the following proportion holds good.

The power exerted by A : power transmitted to C : power expended in circuit :: AF : FE : EA.

Now, suppose a displacement of BC to take place which diminishes the angle ABC—*i.e.*, C is displaced towards M, then evidently E is displaced also towards S; but F is displaced from Q, and therefore AF and FE both become larger.

But AF is proportional to the power exerted by the driver; therefore a greater load is thrown upon the prime mover, and AB will be displaced in the direction of retardation; *i.e.*, the angle CBA will tend to open out again for this reason.

But, further, the power transmitted to C is proportional to FE, and this is increased. The load on C being fixed, and the power being now more than is necessary, the source C will race. Hence, for this reason also, the angle ABC will open out, and regain its original size.

Now, suppose that C is displaced in the direction away from M, which would be the same thing as supposing a retardation of the driver.

In this case AF and FE would both become smaller. Therefore, the prime mover would race, and so AB would advance to close the angle ABC. The source C would have its power cut off, and the load remaining the same, BC would be retarded, and again the angle CBA diminished.

Hence such a position would be one of stable equilibrium of the motion.

As with a magnetic needle in a field of force, there are two positions of equilibrium—one stable, the other unstable; but these points are not (as in the case of the magnetic needle) exactly opposite to one another. On the contrary, the two positions tend to approach each other as the electromotive forces tend to equality, and the angle between their positions is the angle through which the follower may be retarded without permanently upsetting the stability of the motion; but if the displacement exceed this amount, the follower would tend to fall back more and more, and would ultimately come to rest.

There are two positions for E in which no power is transmitted, viz., those which correspond to the points where the two circles SNEQ, BDQA, cut one another, Q and R. If E lies in the section QSR, power is transmitted to the source C; but if E lie on the other portion of the circle it describes, power is exerted by the source C.

For any direction, AE, there exist generally two points where a line drawn from A cuts the circle SNEQ. Let E and E_2 be such a pair of points.

Let C_2 be the position of C corresponding to E_2.

Now, the power exerted by source A for this position will be $\dfrac{AE_2 AF}{2R}$, and that exerted by source C will be

$\dfrac{AE_2 . FE_2}{2R}$. The relation of the latter to the former is $\dfrac{FE_2}{AF}$.

Now, as E_2 moves round from Q to R, on the line QE_2R, the above relation starts from zero, passes by positive values through a maximum value, and ends again at zero. There will therefore be two positions when the relation of the powers has a fixed value, and one of these will be a position of stability, the other unstable for one or other of the sources.

To express the various powers which have been dealt with in terms of the angle ABC, which is the angle by which the difference between the phases of the two electromotive forces falls short of two right angles, proceed as follows: Call this angle θ. Also call the angle BAC, ϕ, and the angle BAD, β.

It is first to be observed that the triangles DEA, BCA, if completed, are similar to each other, since each side of the former bears to some one side of the latter the ratio $\cos \beta$, or geometrically,

$$AE = AC \cos \beta$$
$$ED = CB \cos \beta$$
$$DA = BA \cos \beta.$$

From the triangle ACB, $AC : BC :: \sin \theta : \sin \phi$,

or
$$AC = f \frac{\sin \theta}{\sin \phi},$$

therefore
$$AE = f \frac{\cos \beta \sin \theta}{\sin \phi}.$$

Also from the triangle ABC,
$$\frac{e}{f} = \frac{\sin \overline{\phi + \theta}}{\sin \phi} = \cos \theta + \cot \phi \sin \theta,$$

and from the right-angled triangle AFB
$$AF = AB \cos FAB = e \cos \overline{\phi + \beta}.$$

Hence, AF AE $= ef \cos \beta \sin \theta \dfrac{\cos \overline{\phi + \beta}}{\sin \phi}$

$= ef \cos^2 \beta \sin \theta \cot \phi - ef \cos \beta \sin \theta \sin \beta$

$= ef \cos^2 \beta \left(\dfrac{e}{f} - \cos \theta \right) - ef \cos \beta \sin \beta \sin \theta$

$= e \cos \beta \{ e \cos \beta - f \cos \overline{\theta - \beta} \}.$

Therefore, the power at work in source A being $\dfrac{\text{AF AE}}{2\,\text{R}}$ is equal to

$$\dfrac{e^2 \cos^2 \beta - ef \cos \beta \cos \overline{\theta - \beta}}{2\,\text{R}} \quad \cdot \quad \cdot \quad \cdot \quad \cdot \quad (a).$$

which is in the form desired.

Again, from the triangle CEA,

AE $=$ AC cos CAE $=$ AC cos β,

$\therefore$ AE$^2 =$ AC$^2 \cos^2 \beta$,

and from the triangle ACB

AC$^2 =$ BC$^2 +$ AB$^2 - 2$BC . AB cos ABC,

$= f^2 + e^2 - 2ef \cos \theta.$

Hence, AE$^2 = \cos^2 \beta \{ e^2 + f^2 - 2ef \cos \theta \},$

or the power employed to heat the circuit, being $\dfrac{\text{AE}^2}{2\text{R}}$, is equal to

$$\dfrac{e^2 \cos^2 \beta + f^2 \cos^2 \beta - 2ef \cos \theta \cos^2 \beta}{2\,\text{R}} \quad \cdot \quad \cdot \quad \cdot \quad (\beta).$$

The difference between these two powers is the power transmitted, and therefore doing work upon source C. Hence the power transmitted is equal to

$$\dfrac{ef \cos \beta \cos \overline{\theta + \beta} - f^2 \cos^2 \beta}{2\text{R}} \quad \cdot \quad \cdot \quad \cdot \quad (\gamma).$$

The efficiency of the transmission is found by dividing this expression (γ) by the expression (a) for the power of source A. The efficiency therefore is equal to

$$\frac{f}{e}\left\{\frac{e\cos\overline{\theta+\beta}-f\cos\beta}{e\cos\beta-f\cos\overline{\theta-\beta}}\right\} \quad . \quad . \quad . \quad . \quad (\delta).$$

In Figure 21, let $AB = e$, and the angle $BAD = \beta$. BDA is a circle upon AB as diameter, whose centre is K. Let D be the centre of the circle which is the locus of E of former figures, cutting BDA in Q and R as before, and AD and AD produced in N and P. Join DK and produce it both ways to cut this circle in E_1 and E_2, the point K being in DE_1. Also draw E_4DE_3 parallel to BA, cutting the same circle in E_4 and E_3, making the angle E_3DA equal to β.

The expression for the power of A shows that it is greatest when $\theta = \pi + \beta$, and least when $\theta = \beta$, but it is always positive so long as $e \cos \beta$ is greater than f. When this is not the case it is possible for the source A to have work done upon it.

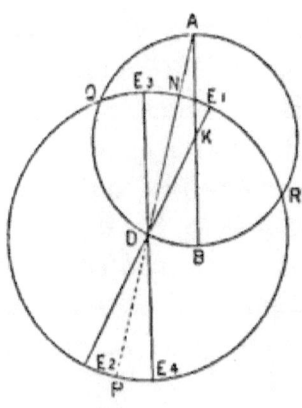

Fig. 21.

Then ADE_3 is the angle giving the value of θ when the power exerted by the source A is least, and ADE_4 (greater than π) is the value of θ for the greatest power of A.

The expression for the transmitted power working on C given by equation (γ) shows that that power is greatest when $\theta = -\beta$, i.e., ADE_1 is the value of θ for this condition, and when $\theta = \pi - \beta$, i.e., ADE_2, the power working on C is negative and a minimum, i.e., C *would have to do its greatest work.*

The circle which E describes is now divided in the important points $QRE_1E_2E_3E_4$, and from the above considerations we can discuss the question of the stability very conveniently. We shall suppose E to move round in the positive direction.

As regards the point E_3, it may be observed, that if DN is sufficiently great, it will lie outside the circle ADB (as in Fig. 22). The condition that it does not do so is expressed by the inequality $\frac{f}{e}$ must be less than $\frac{\cos 2\beta}{\cos \beta}$ which will be supposed to be the case. As E moves from E_1 to Q, there will be transmission of power to C, from its maximum value at E_1 to zero at Q. Hence, as regards C, the motion will be stable, because a retardation of its motion is accompanied by a greater transmission of power to it, and *vice versâ*.

As E moves from Q to E_2, C will do work from zero at Q to its maximum value at E_2.

Hence the motion of C will be stable, because a retardation of its motion is accompanied by a smaller duty, and *vice versâ*.

As E moves from E_2 to R, and finally to E_1, there will be instability as regards C.

Again, as regards the source A. As E moves from E_3 to E_4 through E_2, the power exerted rises from its minimum at E_3 to its maximum at E_4. Hence in this half circle the motion of A will be unstable, because any retardation on its part, implying an advance of E, will be accompanied by a greater duty, and *vice versâ*; so that the displacement will be augmented.

As E moves from E_4 to E_3 through R, the power exerted by A diminishes from its maximum to its minimum value. Hence, in this half circle the motion of A will be stable, because any retardation on its part, implying an advance of E, will be accompanied by a smaller duty, and *vice versâ*; so that the displacement is decreased.

Hence between E_1 and E_3 there is absolute stability for both sources, and between E_2 and E_4 absolute instability.

In the remaining regions there would arise a necessity for constraint in one or other of the sources.

Between E_3 and E_2, the source A would need constraint; between E_4 and E_1, the source C would need it.

If both sources are constrained, a position of E between E_2 and E_4 becomes a possibility.

Under the conditions, therefore, represented by Fig. 21 it would appear impossible to make both sources

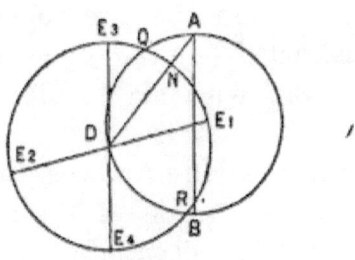

Fig. 22.

do work at the same time without constraint; but this impossibility arises from the fact that E_3 is nearer to N than Q is. When, as in Fig. 22, Q is nearer to N than

E_3 is, *i.e.*, when $\dfrac{f}{e} > \dfrac{\cos 2\beta}{\cos \beta}$, the impossibility is removed, and of the power going to heat the circuit part is provided by each source.

Dr. Hopkinson has denied that two machines can work in series, each doing work; but his reasoning is, in this case, not satisfactory. His statement has, however, been very generally accepted as true by electrical practitioners.

As an instance, suppose $e = 200$ volts,
$f = 150$,,
and $R = 20$ ohms, $\beta = 40°$, $\theta = 30°$.
Then the power exerted by A will be found to be 20·68 watts, and that exerted by C 133·28 watts, the total heating of the circuit being at the rate of 153·96 watts.

And if the prime movers of each machine only yielded the 20·68 and 133·28 watts respectively, the two machines would settle down into stable motion with a phase difference of 150°, and yielding 153·96 watts in rate of heating.

It is perfectly true that one of the machines alone would, *if driven by a sufficiently great power*, yield more power in circuit than the two together; but that is a question distinct from the one whether or not two machines may run with stable motion, in series, both doing work.

As to the transmission of power, its possibility is seen to depend upon the value of the angle E_1DE_3, which is simply 2β. If E is situated at the point N, the efficiency is not at its maximum value, but it has the value $\dfrac{f}{e}$, which is the greatest efficiency in transmission with uniform currents. It would thus appear that transmission by alternating currents does not compare un-

THE TRANSMISSION OF POWER.   93

favourably with that by uniform currents in this respect. To find the position of E which gives the maximum efficiency of transmission, proceed as follows, Fig. 23.

Call the circle described upon AB as diameter, the F circle, since F always lies upon it, and similarly call the circle upon which E always lies, the E circle.

We have to find out the particular position of E upon its circle, which will make the ratio $\frac{FE}{AF}$ as large as possible, or, which is the same thing, that of $\frac{AE}{AF}$ as small as possible.

Let e and f be the two electromotive forces. Take AB equal to e, and cut off from it BM equal to f. Describe the F circle upon AB, and set off BAD as before, an angle having its tangent equal to $\frac{L\pi}{Tr}$.

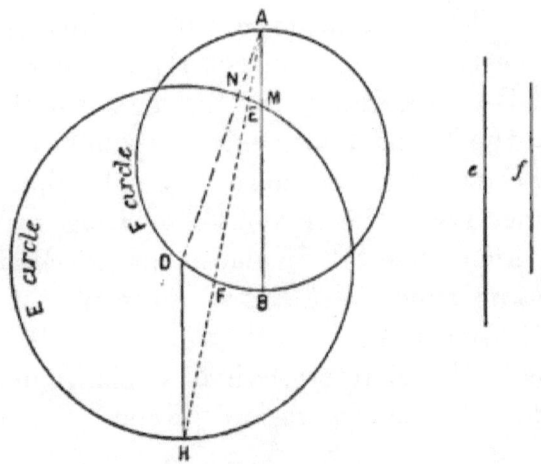

Fig. 23.

D is on the F circle. Draw MN perpendicular to AD, and with centre D at distance DN describe a circle. This is the E circle, for its centre is at D, and its radius = DN = BM cos DAB.

Through D draw the radius DH parallel to AB, H being on the same side of D that M is of A. Join AH by a straight line cutting the E circle in E and the F circle in F. Then these are the particular positions on the circles which make the ratio $\frac{FE}{AF}$ a maximum; *i.e.* $\frac{FE}{AF}$, as here found, is the maximum efficiency. And if DE be joined, the angle ADE will give the phase-difference between the electromotive forces to give this efficiency.

For, in the first place, it is clear that the ratio $\frac{FE}{AF}$ has some maximum (not a minimum) value between the positions where E and F coincide, viz. when the two circles intersect. And because DH is parallel to AB, and AH cuts them, therefore the angle DHE = the angle BAF.

But DH, BA pass through the centres of the two circles respectively; therefore AH, making the same angle with BA as it does with DH, must cut the circumferences of the E and F circles at the same angle. Thus at E and F, the points where AH cuts the E and F circles respectively, the arcs of those circles are parallel. Therefore an elemental displacement of E and F from these positions would result in no alteration of the ratio in which E divides AF.

But this is the characteristic of a maximum or minimum value of such a ratio. Hence the maximum efficiency is $\frac{FE}{AF}$; and it has before been proved that the triangle ADE is similar to one in which the sides homologous to AD, DE are the electromotive forces themselves. Thus, for maximum efficiency, the phases of f must follow the phases of e by an angle equal to

($180° + $ ADE$)$, or, which is the same thing, precede them by ($180° - $ ADE$)$ or ($\pi - $ ADE$)$. This, reduced to time, is

$$\frac{(\pi - \text{ADE})}{2\pi} 2\text{T} \text{ or } \left(1 - \frac{\text{ADE}}{\pi}\right)\text{T}.$$

It is clear that $\dfrac{\text{DN}}{\text{AD}}$ is the ratio of f to e. The maximum efficiency $\dfrac{\text{FE}}{\text{AF}}$ exceeds this; and it is seen to do so in virtue of the existence of a coefficient of self-induction, the absence of which would cause D to coincide with B.

Addendum.—Analytical Expressions.

The value of the maximum efficiency in symbols is

$$\frac{f}{e}\left\{\frac{1 + \dfrac{f}{e}\cos\beta}{\dfrac{f}{e} + \cos\beta}\right\},$$

where $\tan\beta = \dfrac{\text{L}\pi}{\text{T}r}$. The fraction $\dfrac{1 + \dfrac{f}{e}\cos\beta}{\dfrac{f}{e} + \cos\beta}$ can be easily shown to be greater than unity. And the angle ADE may be calculated, if desired, as follows:—

$$\text{ADE} = (2\chi - \beta),$$

where

$$\tan\chi = \frac{\sin\beta}{\cos\beta + \dfrac{f}{e}}.$$

[χ is the angle DAH or BAH.]

Adapting to logarithms.

Let $\dfrac{f}{e} = \cos a,$

$$\tan \chi = \dfrac{\sin \beta}{2 \cos \dfrac{\beta + a}{2} \cos \dfrac{\beta - a}{2}}$$

There may be transmission of power from the source of e to the source of f even when $f > e$, provided that $f \cos \beta$ is not $> e$; as would appear at once from a geometrical construction on the plan given above; and in any case the condition of maximum efficiency is one of stability, as the position of E found by this construction lies within the region for which both machines are in stable motion.

CHAPTER XI.

UPON THE USE OF THE TWO-COIL DYNAMOMETER WITH ALTERNATING CURRENTS.

The dynamometer as usually constructed consists of two coils destined to transmit the same current, and for that purpose placed in series with one another by a maker's connection between the end of the first and the beginning of the second. The user of the instrument has access only to the beginning of the first coil and the end of the second. The torsion to be applied to the suspension of the moving coil to bring it into some constant relative attitude towards the other coil (usually but not necessarily the rectangular position) measures with constant current the square of it, with harmonic current the half of the square, where the maximum value of the current gives the denomination to the current. In this sense I have called the function $\dfrac{I^2}{2}$ of the harmonic current I, the dynamometer reading. But suppose that the employer of a dynamometer has access to the final terminal of the first coil, and to the first terminal of the second coil, as well as to the two usually free terminals; and, further, suppose that he sends one harmonic current through the first coil, and another harmonic current through the second coil, the only ruling condition being that the two currents shall have the same period. What in such a case will be the reading of the instrument? What will be the torsion necessary to bring the coils to the standard relative attitude? It will be as follows: The reading of the dynamometer will measure the quantity $\frac{I_1 I_2}{2} \cos \theta$, when $I_1\, I_2$ are the currents as usually defined by their maximum values, and θ is the angle of the phase difference between the currents, i.e. $\theta = \frac{\pi t}{T}$ where t is the interval of time at which the phases of one current follow the same phases of the other.

The truth of this statement appears at once by applying the first geometrical proposition (see first Article), when the two magnitudes there considered represent $I_1\, I_2$ respectively.

I propose to call the reading of a dynamometer under such circumstances the *Force Reading*, and the function $\frac{I_1 I_2}{2} \cos \theta$ will receive the same name. The propriety of this depends upon the instrument being so graduated that when a direct unit current flows in each coil the angle of the index shall be taken as the unit angle.

This is the same condition under which $\frac{I_2}{2}$ is called the dynamometer reading, under the usual circumstances of connection.

The instrument can be converted into a dynamometer of usual construction by placing a stout wire between the final terminal of the first coil and the first terminal of the second, in fact by making a series arrangement of the coils.

It is clear, therefore, that such an instrument is capable of giving in the first and second place the two dynamometer readings of the two currents —viz., $\frac{I_1^2}{2}$ and $\frac{I_2^2}{2}$, and in the third place $\frac{I_1 I_2}{2} \cos \theta$, or the force reading. A comparison of these three readings will therefore furnish the important angle θ, the angle of phase difference, and, by inference, *all the quantities upon which its value depends.*

Let the three readings be $a_1\ a_2\ a_3$, so that

$$\frac{I_1^2}{2} = a_1, \quad \frac{I_2^2}{2} = a_2, \quad \frac{I_1 I_2}{2} \cos \theta = a_3$$

$$I_1^2 I_2^2 = 4\, a_1 a_2 \therefore \quad \cos \theta = \frac{a_3}{\sqrt{a_1 a_2}}.$$

The final formula, viz., $\cos \theta = \frac{a_3}{\sqrt{a_1 a_2}}$ will not be changed even if the instrument is not graduated as I have supposed, but in any uniform manner, including of course the usual degree graduation.

Before proceeding to examine the application of this method to particular cases, it will be well to point out the following peculiarity. Whereas, when a dynamometer is employed to take a *dynamometer reading*, the twist to be given to the suspension is always in the *same direction*, still, when used for *a force reading*, the twist may have to be applied in either direction. This will merely depend upon the value of $\cos \theta$. If in any case the tendency were in the unusual direction, it may be reversed, should the mechanical arrangement of the suspension or the direction of graduation demand it, by simply making the terminals of one of the coils change places with reference to the main circuit.

The previous chapters have been mainly devoted to

tracing the effect of self and mutual induction and of condensers, local or distributed, upon the values and phases of the resulting currents of the system. All currents in the systems considered have the same period. If, therefore, in the dynamometer we have an instrument capable of measuring the various currents and of detecting their phase differences, it is clear we possess the means of calculating the values of coefficients of self and mutual induction and the capacities of condensers.

To take as an example the case of a harmonic source of E.M.F. operating in a primary coil, which in its turn acts by induction upon a secondary coil. First, let the dynamometer readings for the two currents be taken separately, and suppose them to be a_1 and a_2 respectively. Then let the force reading be taken, and let it be a_3. Then the cosine of the angle of phase difference will be $\dfrac{a_3}{\sqrt{a_1 a_2}}$.

But a reference to Fig. 9, p. 18, shows that this angle is BCF or CEA, whose cotangent is equal to $\dfrac{L'\pi}{rT}$, L' being the coefficient of self-induction of the secondary coil.

But $$\cot^2 \theta = \frac{\cos^2 \theta}{1-\cos^2 \theta}. \quad \therefore \left(\frac{L'\pi}{rT}\right)^2 = \frac{a_3^2}{a_1 a_2 - a_3^2}$$

$$L' = \frac{rT}{\pi} \frac{a_3}{\sqrt{a_1 a_2 - a_3^2}}.$$

Therefore the coefficient of self-induction of the secondary coil is determined, the period and resistance being of course supposed known.

The value of M, the coefficient of mutual induction, may be found without further observation; thus:

$$\frac{CA}{CE} = \sin\theta \quad \text{and} \quad \frac{CE}{CF} = \tan CFE = \frac{M\pi}{RT}.$$

Hence $\quad \dfrac{CA}{CF} = \dfrac{M\pi}{RT} \cdot \sin\theta.$

But $\quad \dfrac{CA}{r} = I_2. \quad \therefore CA = r\sqrt{2\,a_2}.$

Similarly $\dfrac{CF}{R} = I_1. \quad \therefore CF = R\sqrt{2\,a_1}.$

Hence $\quad \dfrac{r}{R}\sqrt{\dfrac{a_2}{a_1}} = \sqrt{\dfrac{a_1\,a_2 - a_3^2}{a_1\,a_2}}\,\dfrac{M\pi}{RT}.$

and therefore $\quad M = \dfrac{rT}{\pi} \cdot \dfrac{a_2}{\sqrt{a_1\,a_2 - a_3^2}}$

or the coefficient of Mutual Induction is determined.

If the coefficient of self-induction in the primary coil is desired, it may be determined by making it temporarily the secondary coil, and pursuing a similar course to that already described. A second determination of M will result from the second series of observations serving as a check upon the former series.

It may be pointed out, in passing, that in the determinations of L′ and M the resistance R in the primary coil need not be known, for its symbol does not occur in the formulæ for those magnitudes. Hence any change in the value of the resistance in the primary coil will not affect the relative values of a_1, a_2, and a_3, upon which M and L′ are calculated, for they are independent of R.

Therefore, the insertion of a resistance in the primary coil may be used to bring all the readings within the range of the instrument, by pairs in turn. Thus, suppose $a_1\ a_2\ a_3$ are in descending order of magnitude. It might happen that when the resistance in the primary is such as to allow a_1 to be read, a_3 might be so small as to be untrustworthy. In this case let a_2 be read. Then take resistance out of the primary circuit. All the readings will go up, a_1 would be off the scale at the high end, but a_3 would reach a reasonable value, and may be compared with a_2. Now a_2 will be found to have gone up in a certain proportion. It will now only be necessary to reduce a_3 in this proportion to obtain a value which, with the former values of a_1 and a_2, will serve for immediate use in the equations.

In investigating the capacity of a condenser by means of an alternating current, we may bridge over a circuit by means of the condenser at points dividing the circuit into two portions, r containing the generator, and R beyond the terminals of the condenser. It may be convenient to avoid a coefficient of self-induction in that part which is remote from the generator, a matter easily accomplished. The coefficient for that part which contains the generator cannot in practice be avoided, but fortunately will not appear in the calculations. For a little consideration will show that if there is no coefficient for the remote section the construction of the diagram upon the rules given for Fig. 15 will coincide with that given for Fig. 11 up to and including the construction of the lines giving the effective electromotive forces in the two sections. So that we need contemplate only the construction of Fig. 11, reproduced so far as is necessary in Fig. 24.

The procedure will be as follows:—

(1) Take the dynamometer readings for the two sections. Let them be a_1 for the generator's section.

a_2 for the remote section.

(2) Take the force reading between the two sections. Let this be a_3.

Then, taking I_1 and I_2 as the currents in the two sections, near and remote, respectively, we must have

$$\frac{OE}{r} = I_1. \quad \therefore \quad OE = r\sqrt{2\,a_1}.$$

$$\frac{EC}{R} = I_2. \quad \therefore \quad EC = R\sqrt{2\,a_2},$$

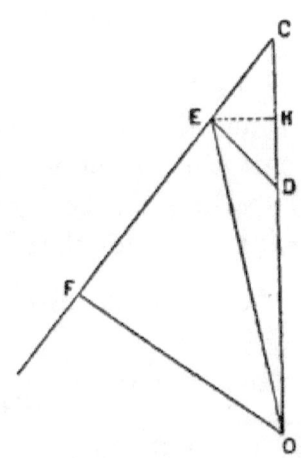

Fig. 24.

and $\dfrac{a_3}{\sqrt{a_1\,a_2}}$ will be magnitude of the cosine of the angle CEO, or the sine of the angle DEO.

Now, from the geometry of the figure,

$$\tan^2 OCF = \frac{OF^2}{FC^2} = \frac{OE^2 - FE^2}{FC^2}.$$

But $$FE = \frac{r}{R} EC = r\sqrt{2\,a_2},$$

and $$FC = \frac{R+r}{R} EC = \overline{R+r}\sqrt{2\,a_2}.$$

Hence $$\tan^2 OCF = \frac{2r^2 a_1 - 2r^2 a_2}{2(R+r)^2 a_2} = \left(\frac{r}{R+r}\right)^2 \frac{a_1 - a_2}{a_2}.$$

Therefore $\tan OCF = \dfrac{r}{R+r} \sqrt{\dfrac{a_1-a_2}{a_2}}$.

But by the construction $\dfrac{C \pi R r}{T(R+r)} = \tan OCF$.

Hence $\dfrac{C \pi R}{T} = \sqrt{\dfrac{a_1-a_2}{a_2}}$.

or C, the capacity of the condenser, $= \dfrac{T}{\pi R} \sqrt{\dfrac{a_1-a_2}{a_2}}$,

thus determined from the dynamometer observations. It will be noticed that this formula does not involve the force reading a_3; but this if taken can be used in the place of a_2, for the values can be shown to be identical.

To prove this point, consider that FEO is the angle of phase difference; therefore its cosine is equal to $\dfrac{a_3}{\sqrt{a_1 a_2}}$.

But cos FEO is also equal to $\dfrac{FE}{EO}$

$$= \dfrac{ECr}{R} \dfrac{1}{EO} = \dfrac{r}{R} \dfrac{R\sqrt{2a_2}}{r\sqrt{2a_1}} = \sqrt{\dfrac{a_2}{a_1}}.$$

Hence $\sqrt{\dfrac{a_2}{a_1}} = \dfrac{a_3}{\sqrt{a_1 a_2}}$,

or $a_2 = a_3$.

This is an important fact in employing the method to measure a condenser, for in practice it would be desirable to have as little self-induction in the remote section of the coil as possible. In this investigation it has been assumed at zero. But the self-induction in the near section has no effect upon the operations. Hence it would be desirable to have the coil in the remote section of no more than a very few turns of wire, while the sensitiveness of the instrument might be furnished by having very many turns in the other coil; i.e., that in the near section. For the same reason it would be well to make the standard position of the coils the rectangular one.

If it be necessary to take the self-induction of the remote section into consideration, we may contemplate the construction of Fig. 15 down to the fixing of the point O.

In this case the observations may be expected to furnish not only the value C of the capacity of the condenser, but also the value of the coefficient of self-induction of the remote section, as in fact they will be found to do. If $a_1\ a_2\ a_3$ are the three observations, we have

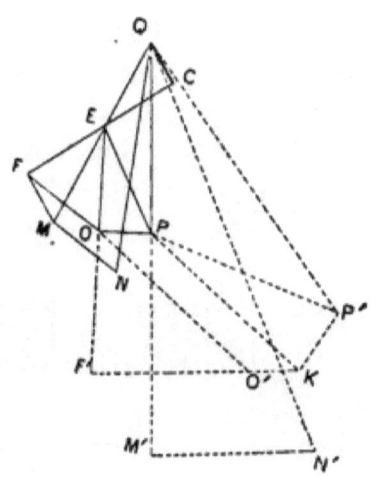

Fig. 15.

$$\text{OE} = r\sqrt{2a_1}, \quad \text{EC} = \text{R}\sqrt{2a_2}.$$

∴ also $\text{EF} = r\sqrt{2a_2}$, and $\cos \text{FEO} = \dfrac{a_3}{\sqrt{a_1 a_2}}.$

For brevity call the angle FEO θ, and the angle

CEQ β. Then also FEM $=\beta$. Because EM is perpendicular to FO,

therefore $\quad$ EF $\cos \beta =$ EO $\cos \overline{\theta - \beta}$,

and $\quad \cos \beta = \dfrac{\text{EO}}{\text{EF}} \cdot \cos \overline{\theta - \beta} = \sqrt{\dfrac{a_1}{a_2}} \cos \overline{\theta - \beta}.$

Hence $\quad \tan \beta = \dfrac{1 - \sqrt{\dfrac{a_1}{a_2}} \cos \theta}{\sqrt{\dfrac{a_1}{a_2}} \sin \theta} = \dfrac{a_2 - a_3}{\sqrt{a_1 a_2 - a_3^2}}.$

But $\quad \tan \beta = \dfrac{L_1 \pi}{TR}. \quad \therefore L_1 = \dfrac{TR}{\pi} \dfrac{a_2 - a_3}{\sqrt{a_1 a_2 - a_3^2}}.$

or the coefficient of self-induction is determined in terms of the dynamometer observations.

This formula also demonstrates the equality of a_2 and a_3 when there is no self-induction.

To find the capacity. From the above value of $\tan \beta$ we may obtain $\quad \cos^2 \beta = \dfrac{a_1 a_2 - a_3^2}{a_2 (a_1 + a_2 - 2a_3)}.$

Further,

$$\tan^2 \text{MQN} = \dfrac{\text{MN}^2}{\text{QM}^2} = \dfrac{R^2 \, \text{FO}^2}{(R+r)^2 \, \text{QE}^2} = \left(\dfrac{R}{R+r}\right)^2 \dfrac{\text{FO}^2}{\text{CE}^2} \cos^2 \beta$$

$$= \left(\dfrac{R}{R+r}\right)^2 \dfrac{\text{FE}^2 + \text{OE}^2 - 2\,\text{FE} \cdot \text{OE} \cdot \cos \theta}{\text{CE}^2} \cdot \cos^2 \beta.$$

$$= \left(\dfrac{R}{R+r}\right)^2 \dfrac{2 r^2 a_2 + 2 r^2 a_1 - 2 r^2 a_3}{2 R^2 a_2} \cdot \cos^2 \beta$$

$$= \left(\dfrac{r}{R+r}\right)^2 \dfrac{a_1 + a_2 - 2 a_3}{a_2} \cdot \dfrac{a_1 a_2 - a_3^2}{a_2 (a_1 + a_2 - 2 a_3)}$$

$$= \left(\dfrac{r}{R+r}\right)^2 \dfrac{a_1 a_2 - a_3^2}{a_2^2}.$$

Hence
$$\tan \text{MQN} = \frac{r}{R+r} \cdot \frac{\sqrt{a_1 a_2 - a_3^2}}{a_2}.$$

But by construction $\tan \text{MQN} = \dfrac{C\pi Rr}{TR+r}.$

Hence
$$C\frac{\pi R}{T} = \frac{\sqrt{a_1 a_2 - a_3^2}}{a_2},$$

and
$$C = \frac{T}{\pi R} \cdot \frac{\sqrt{a_1 a_2 - a_3^2}}{a_2}.$$

Thus the capacity is determined in terms of the dynamometer observations.

Similarly, alternating currents and dynamometers may be employed to investigate the rate at which capacity is distributed along a cable.

The formulæ given above for the values of the coefficient of self-induction, and of the capacity of the condenser, are admirably fitted for calculating their values, to give any desired transformation of the phase and relative value of the currents in the two sections. [*Vide supra* on *Condenser Transformers*.]

Suppose, for instance, that it is desired to maintain equality between the currents, but to put them into quadrature with each other, as in the field of a Tesla motor.

Then (1) $a_1 = a_2$ ⎱ Comprise all the
(2) $a_3 = 0$ ⎰ necessary conditions.

Hence $L_1 = \dfrac{TR}{\pi}$

$$C = \frac{T}{\pi R}.$$

Give the proper values for accomplishing this transformation.

CHAPTER XII.

SILENCE IN A TELEPHONE.

When a conductor subjected to alternating currents is divided between two points into two parallel portions, and when the terminals of a telephone are connected, one to a point in one of those parallel portions, and the other to a point in the other portion, silence may, under certain circumstances, exist; just as when the split conductor is subject to a uniform current, and a galvanometer is substituted for the telephone, no visible motion of the needle may take place.

In either of these cases the failure to detect an electric flow may arise from want of sufficient sensitiveness in the instrument employed—telephone or galvanometer —or to a want of sufficient sensitiveness in the organ of sense employed—ear or eye.

But if in either case no electric flow takes place, then there must be silence in the telephone, and no movement given to the galvanometer needle, however sensitive the instruments, and however quick the senses. Absolute silence in a telephone will therefore exist when its terminals are always kept at equal potentials.

The problem of determining the conditions of absolute silence may be complicated by the existence of coefficients of self-induction in the four portions of the divided part of the conductor, by coefficients of mutual induction, not merely between these four portions, but between them and the rest of the circuit, and by the existence of finite capacity at points of the parallel portions.

I propose merely to deal with the very limited case of the presence of self-induction in each of the four portions of the parallel parts, each section being supposed to be so situated as to suffer and offer no induction from or on any other portion of the system. This case can be easily handled by the geometrical method, and will afford a fair example of its use.

Consider, at first, one of the parallel portions into which the conductor is split, and let $r_1\, r_2$ be the resistances of the two parts into which that portion is divided by the terminal of the telephone; and let $L_1\, L_2$ be the coefficients of self-induction appertaining to each part.

Since there is no current through the telephone, any current through r_1 must exist also through r_2. Hence the currents in these parts must be identical in phase and value. On this account, therefore, the effective electromotive forces through the two parts must be in the same phase, and proportional to the resistances of their respective parts. If, therefore, we take a line PQ, and divide it in A, so that $PA:AQ::r_1:r_2$, PA, AQ may be taken to represent the effective E.M.F.s through r_1 and r_2. At the point A set off the angle PAM, so that $\tan PAM = \dfrac{L_1 \pi}{T r_1}$ where 2T is the period of the alternations.

And at the same point A, but on the side of PQ remote from M, set off the angle QAN, so that $\tan QAN = \dfrac{L_2 \pi}{T r_2}$.

Draw PM, QN, perpendicular to PQ, meeting AM, AN, in M and N respectively.

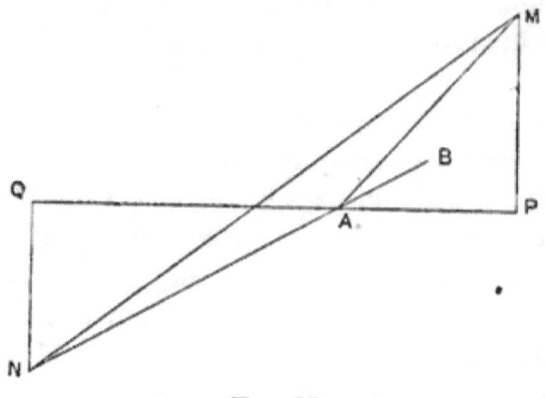

Fig. 25.

Then MA represents the difference of potential of the extremities of r_1.

PM represents the E.M.F. of self-induction in r_1.

AN represents the difference of potential of the extremities of r_2.

NQ represents the E.M.F. of self-induction in r_2.

If MN be joined, MN will represent the difference of potential of the remote extremities of the two parts; *i.e.*, the difference of potential of the points between which the circuit is divided into parallel portions.

But this difference of potential exists for the other parallel portion, the resistances of which are r_3, r_4, let us suppose. If we were to construct a similar diagram to the above for the other parallel, we should arrive at another triangle M'A'N', of which we know that M'N' will represent the same magnitude as MN.

M'A' will represent in phase and magnitude the difference of potential which exists between the second terminal of the telephone, and that end of r_3 which joins on to r_1, which point differs in potential MA from the first terminal of the telephone. If, therefore, the terminals of the telephone are to be always at equal potentials, M'A' must be to MA as M'N' to MN, and the angle NMA must be equal to the angle N'M'A.

The same argument applies to the lines NA, N'A'. *In fact, the triangles must be similar and the conditions of absolute silence in the telephone are expressed in the same way as the conditions of similarity in the triangles,* translating the elements of the triangles into the physical quantities they represent.

The conditions that two triangles are similar to each other are two in number.

Probably the simplest we can select in this case are—

(1) That $\dfrac{MA}{NA}$ shall be equal to $\dfrac{M'A'}{N'A'}$.

(2) That the inclination of MA to NA shall be equal to that of M'A' to N'A'.

Now
$$MA^2 = (PM)^2 + (PA)^2$$
$$\frac{PM}{PA} = \frac{L_1 \pi}{T r_1}$$
$$\therefore MA^2 = PA^2 \left\{ 1 + \left(\frac{L_1 \pi}{T r_1}\right)^2 \right\}$$

Similarly
$$NA^2 = QA^2 \left\{ 1 + \left(\frac{L_2 \pi}{T r_2}\right)^2 \right\}$$

$$\therefore \frac{MA}{NA} = \frac{r_1}{r_2} \cdot \frac{\left\{ 1 + \left(\frac{L_1 \pi}{T r_1}\right)^2 \right\}^{\frac{1}{2}}}{\left\{ 1 + \left(\frac{L_2 \pi}{T r_2}\right)^2 \right\}^{\frac{1}{2}}}, \text{ since } \frac{PA}{QA} = \frac{r_1}{r_2}$$

Similarly
$$\frac{M'A'}{N'A'} = \frac{r_3}{r_4} \cdot \frac{\left\{ 1 + \left(\frac{L_3 \pi}{T r_3}\right)^2 \right\}^{\frac{1}{2}}}{\left\{ 1 + \left(\frac{L_4 \pi}{T r_4}\right)^2 \right\}^{\frac{1}{2}}}$$

in which $L_3 L_4$ are the coefficients of self-induction appertaining to $r_3 r_4$ respectively.

The first condition, therefore, is that

$$\frac{r_1}{r_2} \frac{\left\{1+\left(\frac{L_1\pi}{Tr_1}\right)^2\right\}^{\frac{1}{2}}}{\left\{1+\left(\frac{L_2\pi}{Tr_2}\right)^2\right\}^{\frac{1}{2}}} = \frac{r_3}{r_4} \frac{\left\{1+\left(\frac{L_3\pi}{Tr_3}\right)^2\right\}^{\frac{1}{2}}}{\left\{1+\left(\frac{L_4\pi}{Tr_4}\right)^2\right\}^{\frac{1}{2}}}$$

or

$$r_1^2 r_4^2 \left\{1+\left(\frac{L_1\pi}{Tr_1}\right)^2\right\} \left\{1+\left(\frac{L_4\pi}{Tr_4}\right)^2\right\}$$
$$= r_2^2 r_3^2 \left\{1+\left(\frac{L_2\pi}{Tr_2}\right)^2\right\} \left\{1+\left(\frac{L_3\pi}{Tr_3}\right)^2\right\}. \quad (a)$$

To express the second condition, produce NA towards A to B.

Then the angle BAM is the inclination between MA and AN.

$$\begin{aligned}
\tan \text{BAM} &= \tan\{\text{PAM} - \text{QAN}\} \\
&= \frac{\tan \text{PAN} - \tan \text{QAN}}{1 + \tan \text{PAM} \tan \text{QAN}} \\
&= \frac{\dfrac{L_1\pi}{Tr_1} - \dfrac{L_2\pi}{Tr_2}}{1 + \dfrac{L_1 L_2 \pi^2}{r_1 r_2 T^2}}.
\end{aligned}$$

The corresponding function for the other parallel is

$$\frac{\dfrac{L_3\pi}{Tr_3} - \dfrac{L_4\pi}{Tr_4}}{1 + \dfrac{L_3 L_4 \pi^2}{r_3 r_4 T^2}}$$

Therefore the second condition is that

$$\frac{\dfrac{L_1\pi}{Tr_1} - \dfrac{L_2\pi}{Tr_2}}{1 + \dfrac{L_1 L_2 \pi^2}{r_1 r_2 T^2}} = \frac{\dfrac{L_3\pi}{Tr_3} - \dfrac{L_4\pi}{Tr_4}}{1 + \dfrac{L_3 L_4 \pi^2}{r_3 r_4 T^2}} \quad \cdots \quad (\beta).$$

The conditions (a) and (β) are all that are necessary for absolute silence in the telephone *under the circumstances proposed;* but among the circumstances occurs the period of the alternation, and it is quite possible to have (a) and (β) satisfied for one value of T but not for another. Thus the telephone might be silent for one tone but not for others, as tone depends upon period, and (a) and (β) are only the conditions of the silence in regard to a particular note. It remains to be seen whether it be possible to obtain silence for all notes, one independent of the value of T.

From the construction of the figure, it is clear that if T is very large M and N move to P and Q respectively, and MA becomes coincident with PA, and MN with PQ. Therefore, in order that A may divide PQ in the same ratio for both figures, it will be necessary that $r_1 : r_2 = r_3 : r_4$, or $\dfrac{r_1}{r_2} = \dfrac{r_3}{r_4}$, the usual bridge condition for no current through the galvanometer.

Now, bearing this necessary relation in mind, and noting that MP and NQ are drawn perpendicular to the same straight line PQ, so that $PA : AQ :: r_1 : r_2$, it will be impossible that this relation shall hold good for all straight lines drawn through A unless A lies in the straight line joining N and M.

In this case MAN becomes a straight line, making equal angles with PQ.

Hence
$$\dfrac{L_1 \pi}{r_1 T} = \dfrac{L_2 \pi}{r_2 T}$$

or
$$\dfrac{L_1}{L_2} = \dfrac{r_1}{r_2}$$

and $\dfrac{L_3}{L_4}$ must follow the same rule; *i.e.*—

$$\dfrac{L_3}{L_4} = \dfrac{r_3}{r_4} = \dfrac{r_1}{r_2} = \dfrac{L_1}{L_2}.$$

Al. Cu.

When these conditions are fulfilled, the condition (β) is necessarily satisfied, the inclination of the lines MA, AN, being always zero. Hence these conditions are all that are needed to produce silence in the telephone for all periods of alternations harmonically executed.

CHAPTER XIII.

ON MAGNETIC LAG.

When a conductor or system of conductors is subjected to an alternating current, the magnetic field at any point in the neighbourhood has its sign reversed at every semi-period. If this process involves the continuous rearrangement of anything material, the question naturally arises, Will it not involve a corresponding absorption of energy due to the existence of forces whose direction always opposes the change taking place? If a wind blows over a field of corn alternately from east to west, and from west to east, one can see that a continual generation of heat will take place from the rubbing together of the stalks; and this notwithstanding the natural tendency of each stalk to stand vertically. The forces concerned in this loss are frictional. Does anything of this sort happen with alternating currents of electricity? and does a field of magnetism bear a likeness to a field of corn? In more technical language, does a change in electrokinetic momentum involve a generation of heat, with a corresponding loss of energy in some other form? This question has been answered in the negative by various authorities, and even the Council of the Institution of Civil Engineers has held that view.

I hold the opposite opinion very strongly, on grounds

which I have explained before the Physical Society. The heating of the iron cores of electromagnets subject to alternating currents is too marked a phenomenon to admit of explanation otherwise than in this way. To say that this is due to Foucault currents is only to assign a method to the action, but does not explain it away.

The following paper on Magnetic Lag is reprinted from the *Philosophical Magazine*.

In bringing my views on Transformers before the Physical Society, it is my desire to emphasize :—

(1) How the magnetic lag, if it exist, may be measured by employing dynamometers of low resistance.

(2) That the magnetic lag has a real existence.

(3) That the magnetic lag necessarily accompanies an absorption of work involved in the reversal of polarity in the iron, and how this may be measured.

(4) The points in the general argument where scientific facts are wanting, and the direction which investigation should take to meet this want.

The possibility of the existence of a magnetic lag renders the problem a different one from that of two coils acting and reacting upon themselves by means of mutual and self-induction, whose coefficients, being geometrical, are constant.

For the latter problem I gave in the year 1885 a complete solution, but I pointed out that the completeness of the result rested upon the absence of anything in the nature of hysteresis (a word not then in use) or work done in the field.

The following year Mr. George Forbes, F.R.S., gave what should have been (but for the very poor reporting of the Society of Arts' *Journal*) a solution of the "secondary generator" problem, treating it as a case of two coils, assuming that "the magnetism of the core varies as the sum of the currents in the two coils"; and the same gentleman has treated the subject again in a recent paper before the Society of Telegraph-Engineers and Electricians, in which he makes the same assumption, and says, referring to the harmonic functions which he attributes to the electrical and magnetic quantities involved, that the existence of magnetic hysteresis would cause departure from the harmonic character, but that, being insignificant so long as the magnetic induction in the iron is not high, its consideration may be omitted,—statements which seem rather to evade than to overcome the difficulty.

Mr. Gisbert Kapp, who has done so much good work in the practical development of transformers, also, in my opinion, makes the same assumption,—that the state of magnetization in the core coincides with the magnetic stress resulting from compounding the stresses derived from the two coils.

In the view I shall put forward I shall assume—

(1) That the variations are harmonic.

(2) That the only induction in the secondary coil is derived from the core, and is, therefore, as regards phase, in quadrature with the magnetization. As the current in the secondary coil will be considered as producing one of the components of the stress producing magnetization, itself reacting upon the coil, the necessity of introducing a special E.M.F. of self-induction is obviated.

(3) That each turn in either coil embraces the same number of magnetic lines.

I shall also make use of the following symbols:—

E, the maximum electro-motive force of the machine;
I_1, the maximum value of the current in the primary circuit;

I_2, the maximum value of the current in the secondary circuit;

$\pi - \theta$, the angle of phase-difference between the currents;

m, the number of turns of wire in the primary coil;

n, " " " secondary coil;

ϕ, the angle of magnetic lag;

r_1, the resistance in the primary circuit;

r_2, " " secondary circuit;

$\left.\begin{array}{l}a_1\\a_2\end{array}\right\}$ are the readings of two dynamometers placed respectively in the primary and secondary circuits, their constants being A and B, so that $\dfrac{I_1^2}{2} = A a_1$, $\dfrac{I_2^2}{2} = B a_2$;

a_3 is the reading of a dynamometer, one of whose coils is in the primary, the other in the secondary, circuit. Its constant is C.

M is the maximum magnetization.

The magnetic stress produced by each coil is proportional to the current in that coil multiplied by the number of turns in the coil, and is here taken to be that product, called very often the Ampère-turns. Its maximum value in the primary coil is mI_1, and in the secondary coil it is nI_2.

Now the observations on the dynamometers A and B furnish us with a knowledge of I_1 and I_2 in any case; and m and n are details of the construction of the transformer. Thus we are in possession of the two quantities mI_1 and nI_2.

But the three dynamometer observations enable us to determine the angle of phase-difference between the currents, as I have elsewhere explained:—

$$\cos \theta = \dfrac{C a_3}{\sqrt{A a_1 \, B a_2}} \text{ for } C a_3 = \dfrac{I_1 I_2 \cos \theta}{2}.$$

We are therefore in possession of the two components of the magnetic stress and of the angle between them. Hence we are virtually in possession of the whole magnetic stress and its phase relatively to its components. If the resultant is in quadrature with that component which results from the current in the secondary coil, it is in the same phase as the magnetization, which is in quadrature with that component; but not unless this is the case.

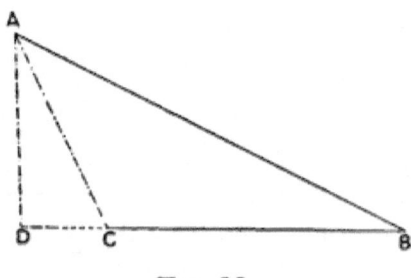

Fig. 26.

Let the line AB represent mI_1 or the magnetic stress in the primary circuit, and let BC represent the magnetic stress in the secondary, and let ABC be the angle θ, found as above. Then AC is the resultant magnetic stress.

But the magnetization is in quadrature with BC. Draw AD at right angles to BC. Then CAD represents the magnetic lag, which is seen to vanish if ACB is a right angle. The condition of the existence of lag is therefore that

$$CB < AB \cos \theta,$$

which in terms of the dynamometer observations is

$$n\sqrt{2\overline{Ba_2}} < m\sqrt{2\overline{Aa_1}} \; \frac{\overline{Ca_3}}{\sqrt{\overline{Aa_1} \cdot \overline{Ba_2}}},$$

or

$$\overline{Ba_2} < \frac{m}{n} \overline{Ca_3}.$$

The observation on the dynamometer in the primary is seen to be eliminated. Thus this question can be tested with two dynamometers only. The amount of lag is represented by the angle CAD. We can easily express its tangent in terms of the three dynamometer observations.

$$\tan \phi = \frac{CD}{DA} = \frac{DB - BC}{AB \sin \theta} = \frac{\dfrac{DB}{AB} - \dfrac{CB}{AB}}{\sqrt{1 - \cos^2 \theta}}$$

$$= \frac{\cos \theta - \dfrac{nI_2}{mI_1}}{\sqrt{1 - \cos^2 \theta}}$$

$$= \frac{\dfrac{Ca_3}{\sqrt{Aa_1\, Ba_2}} - \dfrac{n}{m}\sqrt{\dfrac{Ba_2}{Aa_1}}}{\sqrt{1 - \dfrac{C^2 a_3^2}{Aa_1\, Ba_2}}}$$

$$= \frac{Ca_3 - \dfrac{n}{m} Ba_2}{\sqrt{Aa_1\, Ba_2 - C^2 a_3^2}}.$$

Thus the angle of magnetic lag, if it exist, can be detected with two dynamometers and measured with three.

In dealing yet further with the results furnished by the observations, we must remember that the waxing magnetism has the same inductive effect in producing E.M.F. in each turn of the coils of the two circuits. But we can, from the observation of dynamometer B, say what that E.M.F. per turn is. The whole E.M.F. in the secondary coil is $I_2 r_2$, consequently the E.M.F. per turn is $\dfrac{I_2 r_2}{n}$. Therefore in the primary the total E.M.F. arising from magnetic induction is $m \dfrac{I_2 r_2}{n}$.

The current arising from the same source is $\dfrac{mI_2}{n}\dfrac{r_2}{r_1}$, and the magnetizing stress on this account is $\dfrac{m^2}{n}\dfrac{I_2 r_2}{r_1}$, which must be looked upon as one of the components of the whole magnetizing stress due to the primary current; and this component is in the same phase as the magnetizing stress in the secondary.

Hence, returning to the figure, if we produce CB to F, so that $CB : BF :: nI_2 : \dfrac{m^2}{n}\dfrac{I_2 r_2}{r_1}$,

$$:: 1 : \dfrac{m^2}{n^2}\dfrac{r_2}{r_1}.$$

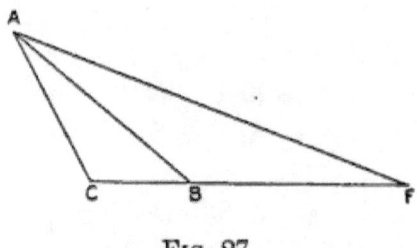

Fig. 27.

FB is one of the components of AB.

The other component (*i.e.*, that arising from the machine's proper electro-motive force) is AF. Hence $AF = m\dfrac{E}{r_1}$.

It follows that the electro-motive force E, which the machine is exerting, may be thus determined by means of the dynamometer observations.

$$AF^2 = AB^2 + BF^2 + 2AB \cdot BF \cos ABC,$$

$$\frac{m^2 E^2}{r_1^2} = m^2 I_1^2 + \left(\frac{m^2}{n^2}\frac{r_2}{r_1}\right)^2 n^2 I_2^2 + 2mI_1 \frac{m^2}{n^2}\frac{r_2}{r_1} nI_2 \cos\theta,$$

$$\therefore E^2 = r_1^2 I_1^2 + r_2^2 I_2^2 \frac{m^2}{n^2} + 2r_1 r_2 \frac{m}{n} I_1 I_2 \cos\theta,$$

$$= r_1^2 2Aa_1 + \frac{m^2}{n^2} r_2^2 2Ba_2 + 4r_1 r_2 \frac{m}{n} Ca_3,$$

$$= 2\left\{r_1^2 Aa_1 + r_2^2 \frac{m^2}{n^2} Ba_2 + 2r_1 r_2 \frac{m}{n} Ca_3\right\}.$$

Another interesting magnitude is AC, or the total impressed magnetic force.

$$AC^2 = AB^2 + BC^2 - 2AB \cdot BC \cos\theta,$$
$$= m^2 I_1^2 + n^2 I_2^2 - 2mn I_1 I_2 \cos\theta,$$
$$= 2m^2 Aa_1 + 2n^2 Ba_2 - 4mn Ca_3,$$
$$= 2\{m^2 Aa_1 + n^2 Ba_2 - 2mn Ca_3\}.$$

By means of this we may calculate what current should be passed through the primary circuit, the secondary being open, to produce the same state in the core.

But perhaps the most interesting point to men of science and to civil engineers is the question of power. We may approach it thus perhaps in the simplest way.

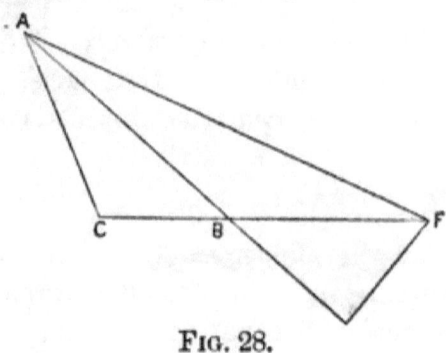

FIG. 28.

By dropping a perpendicular from F upon AB produced, we easily see that
$$AF \cos BAF = BF \cos ABC + AB.$$
Multiplying through by AB we have
$$AF \cdot AB \cos BAF = AB \cdot BF \cos ABC + AB^2.$$

Substituting the electric quantities for the geometrical,
$$\frac{mE}{r_1} mI_1 \cos BAF = mI_1 \frac{m^2 r^2}{n^2 r_1} nI_2 \cos \theta + m^2 I_1^2;$$
multiply through by $\dfrac{r_1}{2m^2}$,
$$\frac{EI_1 \cos BAF}{2} = r_2 \frac{m}{n} \frac{I_1 I_2 \cos \theta}{2} + \frac{r_1 I_1^2}{2}.$$

But the term on the left is the expression for the total power, and those on the right hand may be expressed in terms of the dynamometer observations.

Thus the total power
$$= r_1 A a_1 + r_2 \frac{m}{n} C a_3.$$
The first term here is obviously the power at work heating the primary coil.

$r_2 B a_2$ is as obviously the power heating the secondary coil.

If, therefore, we write the total power
$$= r_1 A a_1 + r_2 B a_2 + r_2 \left\{ \frac{m}{n} C a_3 - B a_2 \right\},$$
we see that the power involved with the magnetic lag is
$$r_2 \left\{ \frac{m}{n} C a_3 - B a_2 \right\},$$
the form showing that it disappears if the lag does so.

Thus we are led to the conclusion that a magnetic lag involves a loss of power, and any loss of power due to molecular action in the core taking place in the course of the alternations of magnetization must necessarily produce lag.

Now, if the changing magnetization does work, it must do it against a force, and this force must be of the character which of itself would produce magnetization; *i.e.*, magnetic force. Just as when a body moving in a medium does work in the medium, it does so by calling into being, or inducing, a force, viz., friction; force being that sort of magnitude which, acting upon a body, produces motion.

By analogy alone, therefore, we may infer that when changing magnetization is a continuous source of absorbed work, the changing magnetization induces what would itself produce magnetization; that is, an induced stress acting in opposition to the direction of the change in the magnetization.

This may perhaps be allowed; but it may be urged that there is plenty of magnetic stress already, impressed by the currents; why should not the changing magnetization work on this? The answer to this objection is, that if there be no other stress but that impressed from outside, then the phase of the magnetization will be in the same phase, and *therefore the increase of magnetization* will be in quadrature with the stress, and hence no work will be continuously absorbed. For though through some phases work may be done in such a case, this is always recoverable and recovered in a complete period, a proposition which I have stated and proved in my papers upon Alternating Currents in 1885.

I apprehend, therefore, that besides the stresses AB BC we have another induced stress in quadrature with the magnetization, because called into being by its increase, and therefore in the same phase as FB or BC.

Let BC, therefore, be produced until it meets in D the line AD drawn at right angles to BC.

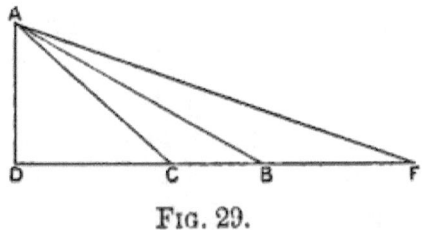

Fig. 29.

Then AD will be the effective magnetic stress; *i.e.*, that magnetic stress which, maintained with a *steady* current, will produce the actual magnetization; hence, if M is the maximum magnetization,

$$\frac{4\pi \text{AD}}{\rho} = \text{M},$$

where ρ is called the magnetic resistance. Thus—

$$\text{AD} = \frac{\text{M}\rho}{4\pi}.$$

If 2T is the period of alternation, $\frac{\pi \text{M}}{\text{T}}$ is the maximum rate of increase of M. If

$$\text{N}\frac{\pi \text{M}}{\text{T}} = \text{DC},$$

N may be called the coefficient of magnetic self-induction.

Under the exigences of a certain temporary nomenclature CD might be called the Foucault-Ampère turns existing in the core itself.

Substituting for M in terms of AD,

$$\text{DC} = \frac{\text{N}\pi}{\text{T}} \frac{4\pi \text{AD}}{\rho},$$

or

$$\frac{\text{DC}}{\text{AD}} = \tan \phi = \left\{ \frac{4\pi^2}{\text{T}\rho} \right\} \text{N}.$$

I have explained above how tan ϕ may be obtained from the dynamometer observations. We may therefore employ this formula for the determination of the value and constancy of N, if we can rely upon the values of T obtained by observation at the time, and of ρ known otherwise.

The constancy of ρ obtains so long as magnetization can keep pace with magnetic stress. These conditions are fairly well understood, and AD must not exceed the stress representing the limiting one. Thus it will be well to know AD.

$$AD = AB \sin \theta = mI_1 \sqrt{1 - \cos^2 \theta},$$

$$= m\sqrt{2Aa_1} \sqrt{1 - \frac{C^2 a_3^2}{Aa_1 \, Ba_2}},$$

$$= \sqrt{2}.m.\frac{\sqrt{Aa_1 \, Ba_2 - C^2 a_3^2}}{\sqrt{Ba_2}}.$$

I think, therefore, experiment should move in the following direction. The constancy or the reverse of N under varying speeds should be first determined by experiments with currents so small or coils so few that the magnetic resistance ρ may be safely assumed constant. For this purpose it would be necessary to employ some speed-indicator.

Professor Forbes, F.R.S., in his paper upon transformers, already quoted, says that the existence of hysteresis would cause a departure from the harmonic character of the motion, but that the effects are small and negligible.

If N, however great, remains constant, the harmonic character of the variation is maintained. But if experience showed that induced stress varies as the square or higher power of the rate of change of magnetization, then indeed a serious modification would take place, and this would be likely if the neighbourhood of saturation were reached.

In the foregoing investigation I have represented *magnetic stress* as of the order *Current*, and the formulæ I have introduced hitherto will bear this convenient form of expression; but, strictly speaking, it is of the order *Field*, and when we wish to settle the dimensions of N, we must take this into account. The M of the work is really of the order Induction or $[l^{-\frac{1}{2}}t^{-1}m^{\frac{1}{2}}]$. Now, the rate of increase of Induction multiplied into N is equal to Field, or $[l^{-\frac{1}{2}}t^{-1}m^{\frac{1}{2}}]$;

$$\therefore N \frac{[l^{-\frac{1}{2}}t^{-1}m^{\frac{1}{2}}]}{t} = [l^{-\frac{1}{2}}t^{-1}m^{\frac{1}{2}}],$$

$$\therefore N = [t].*$$

The actual experiments, which I now bring to the notice of the Society, were carried out at the works of Messrs. Nalder, Brothers & Co., of Westminster, and I am greatly indebted to these gentlemen themselves, and to Mr. Crawley and Mr. Mott, for their assistance in making them, for they freely placed their steam-power, their electrical power, and their intellectual power at my disposal.

These experiments had no other object than to test the question of the existence of magnetic lag by dynamometers, and to measure the angle of lag.

The machine employed and the transformer were of the Kapp forms. The number of turns of wire in the

---

* Or $[t\mu^{-1}]$ according to Prof. A. W. Rücker's more complete formulation. Compare with this the coefficient of Electric Self-Induction $[l\mu]$.

two coils had been ascertained for me by Mr. Crawley. They were 100 and 12 respectively.

Considering the considerable differences in the relations of the currents, the constancy of the angle of lag appears to me to point to a simple law connecting it with the magnetization.

There was no very accurate speed-indicator employed, and the speed was approximately the same in the first six experiments. In the seventh experiment the speed was purposely much diminished, by about one third of that in the former cases, and in this experiment alone does the lag appear at less than 5°.

No. of Experiment.	$Aa_1 = \dfrac{I_1^2}{2}$	$Ba_2 = \dfrac{I_2^2}{2}$	Ca_3.	$\dfrac{m}{n} Ca_3$.	θ.	ϕ.	$\sqrt{Aa_1} \sin \theta$.
1...	33·29	52·65	10·37	86·46	75° 40′	5° 43′	5·590
2...	34·43	59·74	11·50	95·83	75° 19′	5° 38′	5·676
3...	37·09	92·14	17·34	144·50	72° 45′	6° 25′	5·816
4...	70·38	86·06	17·43	145·22	77° 4′	5° 21′	8·176
5...	81·17	81·00	17·21	143·42	77° 45′	5° 24′	8·804
6...	84·97	29·03	7·83	65·26	80° 56′	5° 4′	9·103
7...	8·21	101·59	14·31	119·26	60° 17′	4° 50′	2·488

No. of Experiment.	r_1.	r_2.	$\dfrac{I_1^2}{2} r_1$.	$\dfrac{I_2^2}{2} r_2$.	Power of internal magnetic work = H.	$\dfrac{H}{Aa_1 \sin^2 \theta}$
1...	2·01	2·51	66·91	132·15	84·96	2·72
2...	,,	2·36	69·20	140·99	85·17	2·64
3...	,,	1·96	74·55	180·59	102·63	3·03
4...	,,	...		141·46		
5...	,,	2·87	163·15	232·47	179·15	2·31
6...	,,	4·77	170·79	138·47	172·82	2·09
7...	,,	0·72	16·50	73·14	12·73	2·06

Table A.

θ	$\dfrac{\cosh\theta + \cos\theta}{2}$	$\dfrac{\cosh\theta - \cos\theta}{2}$	$\cosh\theta$	$\cos\theta$
0·0	1·	·0	1·	1·
·01	1·	·0000500	1·0000500	·9999500
·02	1·	·0002000	1·0002000	·9998000
·03	1·	·0004500	1·0004500	·9995500
·04	1·0000001	·0008000	1·0008000	·9992001
·05	1·0000003	·0012500	1·0012503	·9987503
·06	1·0000005	·0018000	1·0018005	·9982005
·07	1·0000010	·0024500	1·0024510	·9975510
·08	1·0000017	·0032000	1·0032017	·9968017
·09	1·0000027	·0040500	1·0040527	·9959527
·10	1·0000042	·0050000	1·0050042	·9950042
·2	1·0000667	·0200001	1·0200668	·9800666
·3	1·0003375	·0450010	1·0453385	·9553365
·4	1·0010667	·0800060	1·0810724	·9210607
·5	1·0026043	·1250217	1·1276260	·8775826
·6	1·0054004	·1800648	1·1854652	·8253356
·7	1·0100056	·2451634	1·2551690	·7648422
·8	1·0170708	·3203641	1·3374349	·6967067
·9	1·0273482	·4057382	1·4330864	·6216100
1·00	1·0416915	·5013892	1·5430807	·5403023
1·1	1·0610573	·6074612	1·6685185	·4535961
1·2	1·0865066	·7241489	1·8106555	·3623577
1·3	1·1192066	·8517078	1·9709144	·2674988
1·4	1·1604328	·9904656	2·1508984	·1699672
1·5	1·2115614	1·1408362	2·3513978	·0707252
1·6	1·2741325	1·3033320	2·5774645	−·0291995
1·7	1·3497354	1·4785800	2·8283154	−·1288446
1·8	1·4401355	1·6673376	3·1074731	−·2272021
1·9	1·5472209	1·8705105	3·4177314	−·3232896
2·0	1·6730244	2·0891713	3·7621957	−·4161469
2·1	1·8197334	2·3245796	4·1443130	−·5048462
2·2	1·9897036	2·5782048	4·5679084	−·5885012
2·3	2·1854716	2·8517475	5·0372191	−·6662759
2·4	2·409776	3·147170	5·556947	−·7373937
2·5	2·665573	3·466717	6·132290	−·8011436
3·0	4·538775	5·528771	10·06766	−·989996
4·0	13·15412	13·98094	27·30824	−·82682

www.ingramcontent.com/pod-product-compliance
Lightning Source LLC
Chambersburg PA
CBHW020103170426
43199CB00009B/372